The Structures and Properties of Solids
a series of student texts

General Editor:
Professor Bryan R. Coles

The Structures and Properties of Solids 2

The Crystal Structure of Solids

P. J. Brown, M.A., Ph.D.
Fellow and Lecturer in Physics, Newnham College Cambridge
Assistant Director of Research in the Cavendish Laboratory

J. B. Forsyth, M.A., Ph.D.
Neutron Beam Research Unit
Science Research Council
Rutherford High Energy Laboratory, Chilton, Berkshire

Edward Arnold

© P J Brown and J B Forsyth 1973

First published 1973 by Edward Arnold (Publishers) Limited
25 Hill Street, London W1X 8LL

Boards Edition ISBN: 07131 2387 7
Limp Edition ISBN: 07131 2388 5

All rights reserved. No part of this publication may be reproduced, stored in a retrieval system, or transmitted in any form or by any means, electronic, mechanical, photocopying, recording or otherwise, without the prior permission of Edward Arnold (Publishers) Limited.

548.84 BRO

67812

Printed in Great Britain by William Clowes & Sons Ltd., London, Colchester and Beccles

General Editor's Preface

Most of the solids with which the solid state physicist or materials scientist are concerned are crystalline, sometimes single crystals, more often polycrystalline aggregates, and it is vital to an understanding of their behaviour to possess an understanding of the atomic arrangements and consequent crystal symmetry in these materials. Most textbooks of crystallography, however, provide more detail of crystallographic techniques than the undergraduate or post-graduate student of materials science or physics normally requires. On the other hand they frequently give only a very brief account of the crystal structures of particular solids and the physical and chemical principles underlying the appearance of these structures.

The present book, one of a series designed to cover the central topics in solid state science, aims to provide the basic understanding of crystal structures, their examination by diffraction techniques and the influences that govern their form, required by a scientist concerned with solids at the atomic level. The careful account of diffraction of different types of radiation provides both the basic theory of crystallographic techniques and an account valuable in its own right of the interaction of radiation with static crystals. (The use of scattering studies to provide information about *The Dynamics of Atoms in Crystals* is the subject of the next text in the series by Professor W. Cochran). The discussion of the arrangements of atoms in real crystals requires for completeness the brief discussion given of the behaviour of electrons in crystals of different types; this topic will be the subject of a separate text to appear in the series—*The Electronic Structures of Solids* by B. R. Coles and A. D. Caplin.

Drs Brown and Forsyth bring to this book their experience of many types of crystallographic investigation, and an understanding, from their own experience, of the needs in this field of many different types of scientist.

Imperial College,
London,
1973.

BRC

Preface

This book gives an introductory account of the crystal structures found in solid materials. It is intended for the student of solid state physics or materials science who wishes to understand and make use of crystallographic techniques and the results which have been obtained by using them.

In the first chapter the ideas and nomenclature which form the foundations for the description of crystals and their structures are laid. This is followed by an account of the exceedingly powerful diffraction techniques which, during the past sixty years, have made possible the study of crystal structure on an atomic scale and have led to our knowledge of the bewildering variety of different structures known today. First, the production and properties of the three most commonly used radiations, X-rays, neutrons and electrons, are described. The next two chapters develop the theory of the scattering of radiation by crystalline materials; in the first the emphasis is placed on weak scattering by a general periodic potential and in the second the special features and areas of applicability of the scattering of X-rays, neutrons and electrons are considered. A final chapter on diffraction introduces a few of the experimental techniques which may be of use to the non-specialist crystallographer, such as the all-important powder technique and methods for the determination of the orientation of single crystals.

Having thus introduced the major techniques for the study of crystals, the rest of the book is devoted to the principles underlying the stability of structures. An attempt has been made to avoid a rigid classification of structures by 'bond type', but rather to stress the relationships as opposed to the distinctions between the myriad crystal structures found in nature. The crystal structures of the elements are considered in some detail, since these illustrate in the simplest way most of the important principles of structure building. The necessary extra complications which occur in crystals containing more than a single atomic specie are covered in the next two chapters, the first on polar structures dealing with chemically dissimilar atoms and the second on binary alloys which illustrates the effects of size differences. No treatment of the structures of organic compounds is appropriate since they add little to our understanding of structural principles.

Finally, some guidance is given to the solid state physicist or materials scientist encountering a new material about the ways in which he may establish whether its crystal structure is already known and, if not, whether it is isomorphous with some known structure type.

Cambridge and Chilton
1973

PJB
JBF

Contents

3 GENERAL THEORY OF CRYSTAL DIFFRACTION

4 THE SCATTERING OF X-RAYS, ELECTRONS AND NEUTRONS

5 EXPERIMENTAL STUDY OF DIFFRACTION BY CRYSTALS

1

Crystallographic Geometry and Symmetry

1.1 Introduction: the crystalline state

Any study of solids without a knowledge of the details of the atomic arrangements must of necessity be of a superficial nature. The science of crystallography is concerned with the theory and techniques by which these arrangements, the crystal structures, are established. The great majority of all solid materials are crystalline, that is the atoms of which they are composed are arranged in a highly regular way. It is this regularity, together with the attendant symmetry, that characterises the crystalline state. Only infrequently do the conditions of crystal growth favour the production of the large plane faces which are popularly associated with 'crystals.'

The purpose of this book is to introduce the most important crystallographic techniques of general use to the student of the solid state and to describe a number of crystal structures, chosen to illustrate the interplay of factors which determine the atomic arrangement. The fundamental technique for structure determination is that of diffraction. A brief account will be given of the production and properties of X-rays, neutrons and electrons with wavelengths in the useful range and both theoretical and experimental aspects of their diffraction will be treated.

The science of crystallography had its origins in the study of the external morphology of crystalline minerals. The flat faces exhibited by these materials became the subject of quantitative study and it was found that any particular mineral could be characterized by its interfacial angles as measured by a contact or optical goniometer. These measurements led to a study of the symmetry and geometry of crystals which has been carried over usefully to the description of their internal structures.

1.2 Crystal symmetry and the crystal lattice

Before considering the external symmetry of crystals which is best described by means of the stereographic projection (§1.4) it seems more profitable to consider the nature of the atomic arrangement in crystalline solids, its symmetry and its periodicity, since these properties give rise to the external morphology of crystals and to anisotropy in their physical properties.

It is convenient to describe the symmetry of a crystal structure by means of symmetry elements. A *symmetry element* is an operation which leaves the pattern of the atomic arrangement unchanged. Figure 1.1 shows an array of circles in contact: it can easily be seen that the operation of rotations through $2\pi/6$ about an axis perpendicular to the plane of the paper and passing through the point A leaves the pattern unchanged. This axis is called a sixfold *rotation axis* or hexad. Similarly, there is a threefold rotation axis or triad through B. Another common symmetry operation is that of reflection and the symmetry element to which it corresponds is a *mirror plane.* There are mirror planes perpendicular to all the solid lines in Fig. 1.1.

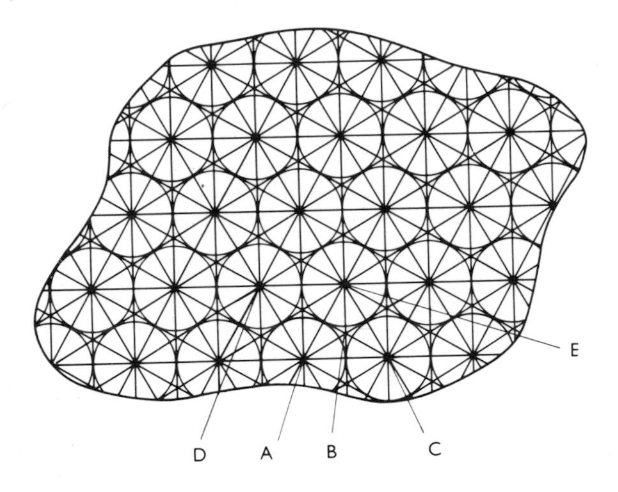

Figure 1.1 Symmetry operators in part of an infinite two-dimensional array. Hexads pass through the points A and triads through the points B. There are mirror planes parallel to all the solid lines. Translations through the vector distances AC, AD, AE, CA, DA and EA leave the pattern unchanged.

If we ignore the boundaries of the crystal, there is a further class of operations which leave the pattern unchanged, namely translations; for example, the movement of the crystal through the vector distance AC (Fig. 1.1). In general, if we consider any point $\boldsymbol{r}$ in an ideal crystal then there are other points $\boldsymbol{r}'$ throughout the crystal which have identical environments. Here

$$\boldsymbol{r}' = \boldsymbol{r} + n_1\boldsymbol{a} + n_2\boldsymbol{b} + n_3\boldsymbol{c} \tag{1.1}$$

and n_1, n_2, n_3 are arbitrary integers. The fundamental translation vectors $\boldsymbol{a}$, $\boldsymbol{b}$ and $\boldsymbol{c}$ define the crystal axes and the displacement $n_1\boldsymbol{a} + n_2\boldsymbol{b} + n_3\boldsymbol{c}$ is a *translation operation.*

A *unit cell* of a crystal is defined as any polyhedron with the following properties:

a) None of the above translation operations, other than the identity operation $n_1 = n_2 = n_3 = 0$, results in a translated parallelepiped which overlaps the original one.

b) The complete set of parallelepipeds generated from it by all the above translation operations covers all points in space.

A convenient way to select a unit cell is to associate an arbitrary point with its origin: the infinite array of points $\boldsymbol{r}'$ described by Equ. 1.1 with $\boldsymbol{r} = 0$ then constitute the crystal *lattice.* In some simple structures the origin is chosen to be coincident with the centre of an atom, but this is by no means essential. *The essential property of a lattice is that every lattice point has the same environment in the same orientation.* A particular lattice may be described in any number of ways by different choices of the basis vectors $\boldsymbol{a}$, $\boldsymbol{b}$ and $\boldsymbol{c}$, which are inclined to each other at angles α, β and γ. The most general shape of the unit cell is a parallelepiped, with sides of lengths a, b and c respectively, and in which $a \neq b \neq c$ and $\alpha \neq \beta \neq \gamma$. The external morphology of the crystal may, however, show the following additional symmetry elements:

a) Rotations of $2\pi/n$ about axes through the origin, where $n = 1, 2, 3.$ 4 or 6.

b) Reflections in planes containing the origin.

c) A *centre of inversion* which takes $\boldsymbol{r}$ into $-\boldsymbol{r}$ accompanied by a change in hand.

The arrangement of lattice points must possess all the symmetry of the external morphology and it is appropriate to choose a unit cell which reflects this symmetry. For example, the presence of a fourfold rotation axis (tetrad) enables a cell which is a tetragonal prism to be identified.

1.3 The Bravais lattices and the seven crystal systems

As long ago as 1848, Bravais was able to show that there are only 14 space lattices and their unit cells are illustrated in Fig. 1.2.

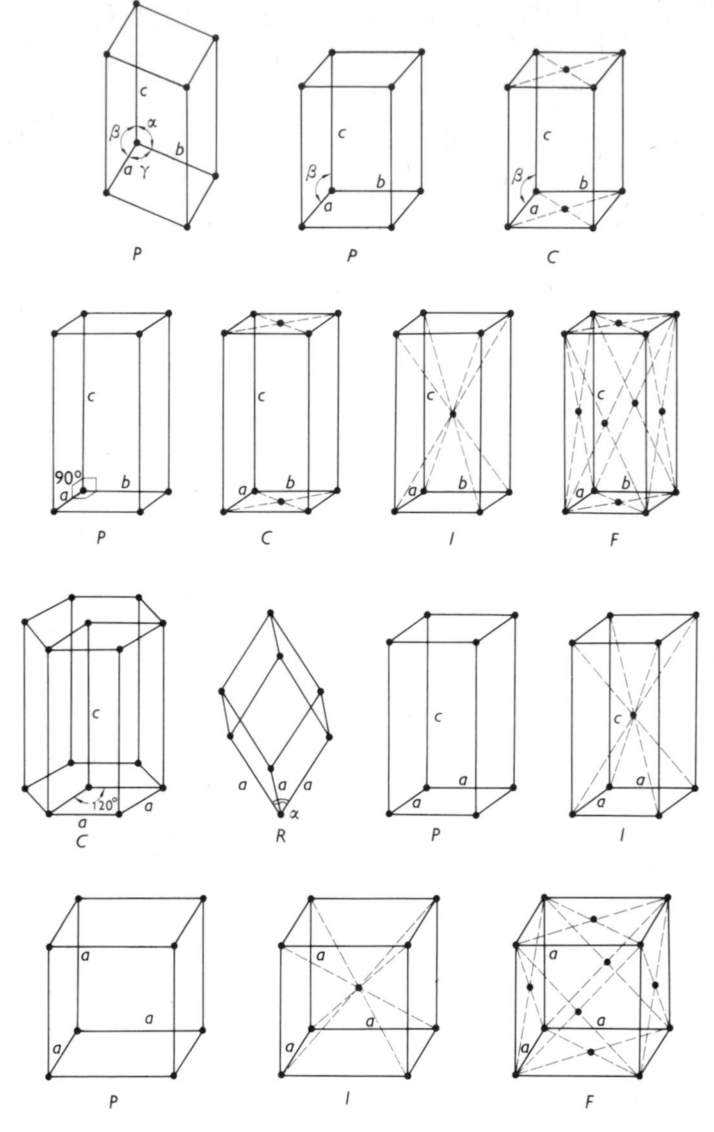

Figure 1.2 The Bravais lattices.

There are seven different shapes of unit cell each containing the equivalent of one lattice point. Each of these cells exhibits the highest symmetry possible (*holosymmetry*) in one of the seven *crystal systems*.

The cells selected for the remaining lattices contain an additional lattice point at the centre (I), in the middle of each face (F), or in the middle of a pair of faces (C(A or B)); these are therefore not true unit cells, but they were chosen by Bravais in order to bring out more clearly the relationship between the lattice and the crystal symmetry. The other possible combinations of adding additional lattice points to the seven primitive (P) Bravais lattices result in arrangements which are either (a) no longer lattices, or (b) lattices, but ones with a different symmetry to the original primitive lattice, or (c) lattices with the same symmetry as the original lattice but which may be described by one of the 14 preferred Bravais lattices after a change in axes allowed by the particular lattice symmetry.

Examples of (c) above are illustrated in Fig. 1.3 in which we show that in

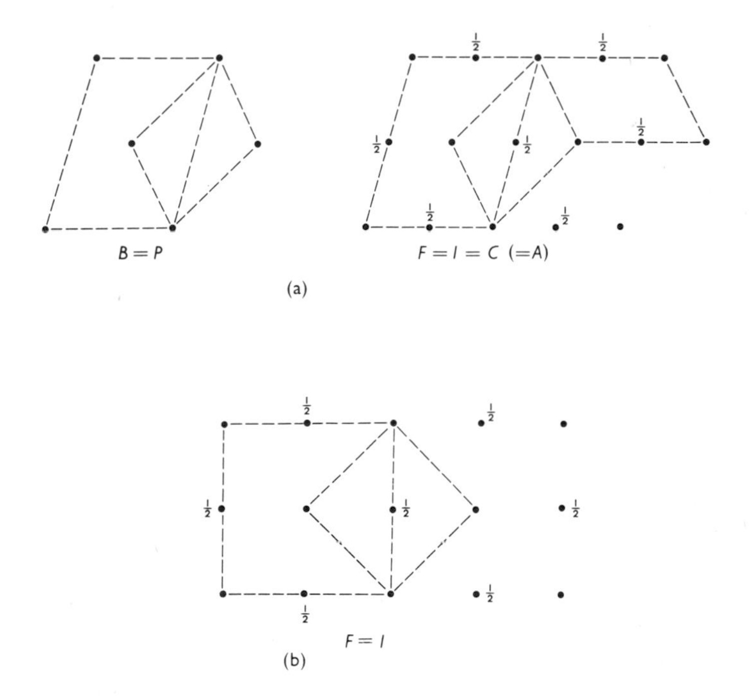

Figure 1.3 Equivalence between Bravais lattices (a) in the monoclinic system and (b) in the tetragonal system.

the monoclinic system $B = P$ and $F = I = C(= A)$ with a fresh choice of $\boldsymbol{a}$ and $\boldsymbol{c}$ axes. In the tetragonal system $F = I$, by re-defining the axes perpendicular to the tetrad. Figure 1.3 introduces the method of representing a set of lattice points or atomic positions by their projected positions on a convenient plane which will be used frequently throughout this book. The distance of any point from the plane of projection is given as a fraction of the repeat distance parallel to the projection axis.

The characteristic symmetry and unit cell dimensions for each of the crystal systems are listed in Table 1.1.

Table 1.1 The seven crystal systems

System	Unit cell dimensions		Essential symmetry
Triclinic	$a \neq b \neq c$	$\alpha \neq \beta \neq \gamma$	None
Monoclinic	$a \neq b \neq c$	$\alpha = \gamma = 90°$	One twofold axis of rotation or inversion conventionally chosen parallel to ***b*** axis.
Orthorhombic	$a \neq b \neq c$	$\alpha = \beta = \gamma = 90°$	Three mutually perpendicular twofold axes of rotation or inversion parallel to ***a***, ***b*** and ***c***.
Tetragonal	$a = b \neq c$	$\alpha = \beta = \gamma = 90°$	One fourfold axis of rotation or inversion parallel to *c*.
Hexagonal	$a = b \neq c$	$\alpha = \beta = 90°$ $\gamma = 120°$	One sixfold axis of rotation or inversion parallel to *c*.
Trigonal or Rhombohedral	$a = b = c$	$\alpha = \beta = \gamma < 120°$ and $\neq 90°$	One threefold axis of rotation or inversion parallel to $\boldsymbol{a} + \boldsymbol{b} + \boldsymbol{c}$.
Cubic	$a = b = c$	$\alpha = \beta = \gamma = 90°$	Four threefold axes parallel to the body diagonals of the unit cell.

1.4 The stereographic projection

At this point we shall interrupt our consideration of the internal symmetry of crystals to introduce the stereographic projection. It provides a convenient way of representing, in two dimensions, the three-dimensional relationships between symmetry elements or crystal faces.

The use of perspective drawings to illustrate the faces exhibited by a crystal is unsatisfactory: they are difficult to construct and do not yield quantitative information easily. In the stereographic projection a sphere is imagined to surround the crystal and from the centre of the sphere normals are drawn to the crystal planes (Fig. 1.4a). The point at which each normal touches the sphere is then projected back through the equatorial plane, which is the plane of projection, to the pole of the lower hemisphere as in Fig. 1.4b. The points where these lines cut the equatorial plane are the *stereographic poles* of the corresponding faces. The stereographic poles of faces whose normals cut the sphere in the lower hemisphere will project outside the projection of the equator (or *primitive circle*): they may also be projected inside this circle by using the upper pole as the pole of projection. In this case, the poles are shown as open circles to distinguish them from poles projected from the upper hemisphere which are shown as dots. A completed stereogram is shown in Fig. 1.4c; the curves connecting the poles are *great circles* (i.e. circles whose planes pass through the centre of the original sphere) or *zones*. A set of faces whose normals are co-planar are said to lie in a zone and the edges formed by their intersections are all parallel to a single direction which is called the *zone axis*. It is a property of the stereographic projection that all zones project as circles on the stereogram and pass through diametrically opposite points on the primitive circle.

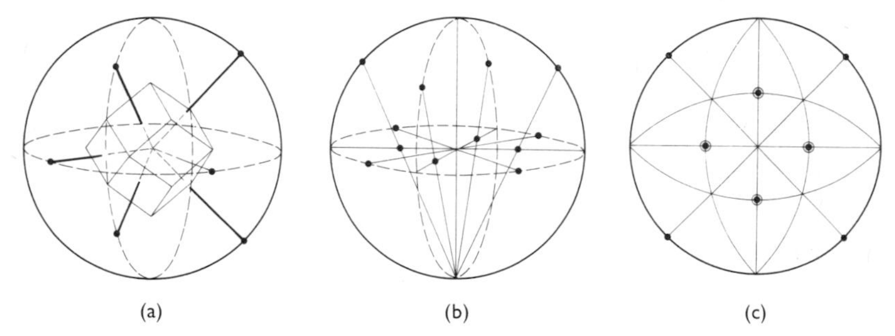

Figure 1.4 The stereographic projection of a rhombic dodecahedron.

A *small circle* or projection sphere is the locus of points which are at equal angular distances from a point on the sphere. Small circles project as circles, but in general the centre of the small circle will not project as the centre of the circle on the stereogram (Fig. 1.5).

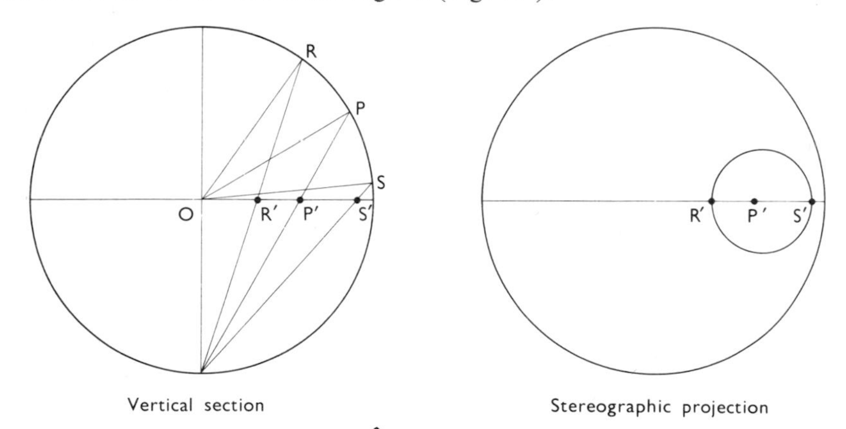

Figure 1.5 A small circle of angle $R\widehat{O}P$ about P projects as a circle with diameter R′P′S′, but with the projection of P at P′ no longer at its centre.

An aid to the use of the stereographic projection is provided by the *stereographic net* which is illustrated in Fig. 1.6. It consists of a series of great circles drawn at a suitable angular interval between two diametrically

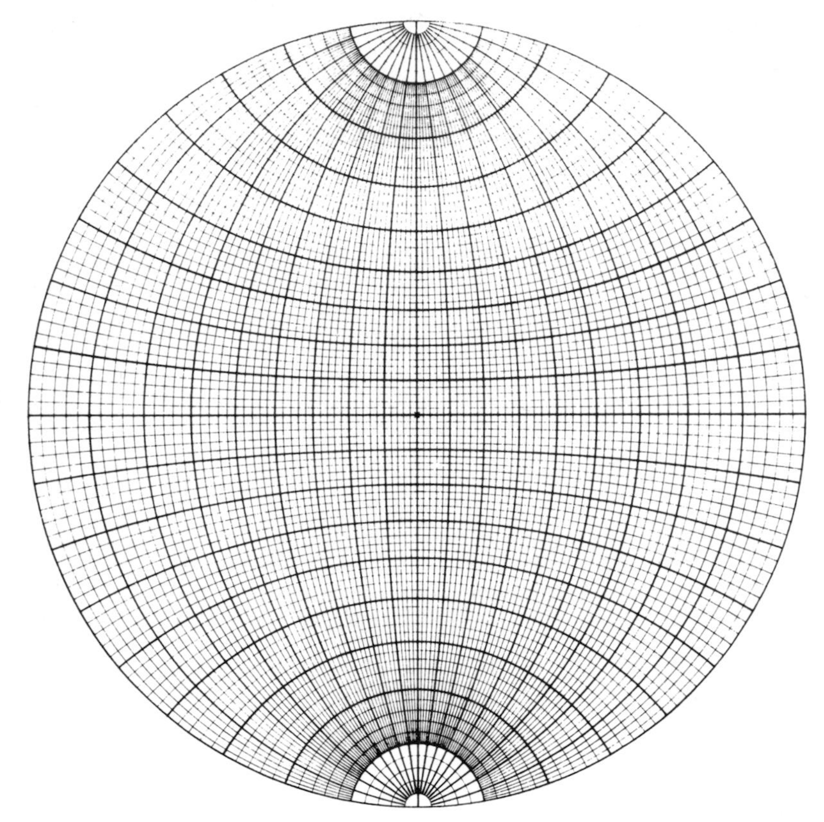

Figure 1.6 The Wulff net drawn at 2° intervals.

opposed points on the primitive circle; superimposed on this is a series of small circles drawn about these two diametrically opposite points. A commonly available net is based on a projection sphere of 2·5 in radius with angular intervals of 2°. By rotating the net about its centre, all possible great circles can be obtained together with all the small circles about points on the primitive circle. Should a small circle be required about a point lying within the primitive circle the net can be used to locate the two points which lie on the diameter drawn through the pole and which are at the appropriate angular distance from it. The projection of the small circle can be constructed with compasses so that the two points lie on either end of its diameter.

The angular distance between two face normals in a stereogram can be determined by rotating the stereographic net so that their poles both lie on the same great circle: the distance between them is then obtained from the intersections of the small circles with this particular zone.

The zone axis or pole of a great circle can easily be located on the stereogram since it is 90° from all points in the zone (Fig. 1.7). The angle

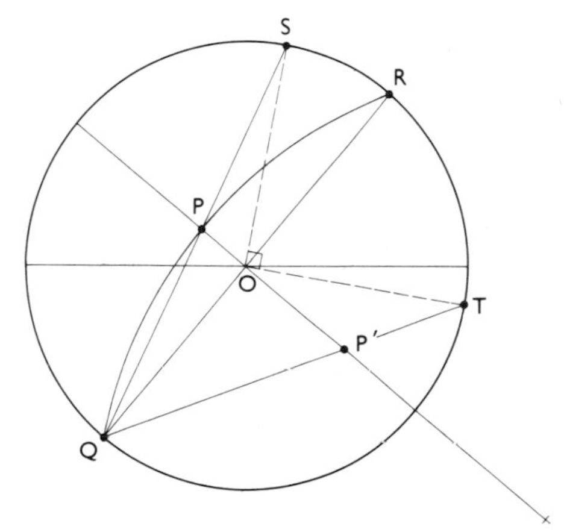

Figure 1.7 Construction to locate the pole P′ of the great circle QPR. Project P on to the primitive from Q to S; locate T such that $S\hat{O}T$ is a right angle. Re-project T from Q to give the pole at P′.

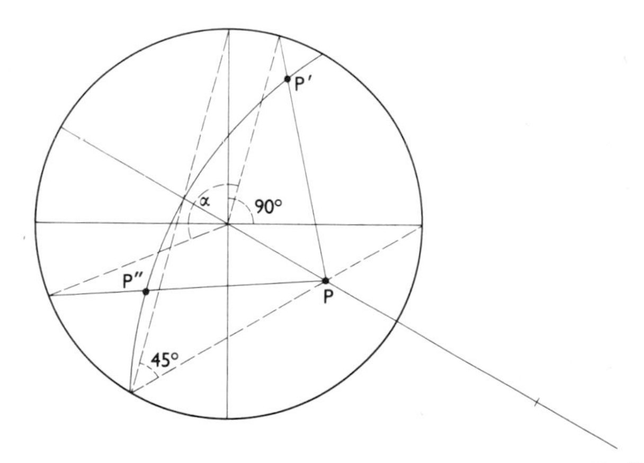

Figure 1.8 Construction to determine the angle between two poles P′, P″.
P is the pole of the great circle joining P′ and P″ which can be located using the construction illustrated in Fig. 1.4.

between two zones is the angle between their zone axes, which may be found in exactly the same way as the angle between two face normals: a graphical method for determining such an angle is given in Fig. 1.8.

1.5 Point groups

The complete set of symmetry operators in a crystal is called its *space group.* The group of operations which is obtained by setting all translations of the space group elements equal to zero is called the *point group* of the crystal, and it is this point group which controls its external morphology. There are 32 such ‘crystallographic’ point groups which are consistent with translational symmetry and a total of 230 space groups. (In enumerating the 32 point groups appropriate to crystal symmetry, it is convenient to start with the groups containing the minimum symmetry appropriate to the seven crystal systems and listed in Table 1.1.) The symmetry of a point group is most conveniently given either as a symbol or as a stereogram (see § 1.4) showing the point group of symmetry elements. Unfortunately there are two systems of symbols in current use: those due to Schönflies and to Hermann-Mauguin. The latter have been preferred by the International Union of Crystallography and are used in all its compilations; the former are frequently used in works on group theory. We shall adopt the International symbols, which are listed together with their meanings in Table 1.2, since their extension to the description of space groups gives

Table 1.2 Point group symbols

Symbol	Meaning
X or $\bar{X}$	Rotation or inverse axis respectively.
$X2$ or $\bar{X}2$	Subsidiary diad or diads exist perpendicular to X or $\bar{X}$.
Xm or $\bar{X}m$	The m denotes a symmetry plane or planes *containing* the axis X or $\bar{X}$.
X/m	Symmetry plane *perpendicular* to the axis X.
X/mm	A combination of Xm and X/m.
mm	Two mutually perpendicular symmetry planes: these intersect in a diad and hence the symbols $2mm$ and $mm2$ are sometimes used.
$\bar{X}2m$ or $\bar{X}m2$	The symmetry planes contain the axis $\bar{X}$ whereas the diads are perpendicular to $\bar{X}$.
23, $\bar{2}3$, $m3$	Cubic system symbols always contain 3 as the second symbol denoting the four triads.
$\bar{4}3m$, 432	The first symbol is m if there are symmetry planes perpendicular to ⟨100⟩, otherwise it denotes the rotation axes parallel to ⟨100⟩.
$m3m$	A third symbol refers to the ⟨110⟩ symmetry: a diad parallel or a symmetry plane perpendicular to these directions.

much more information than that obtainable from the Schönflies notation. For convenience, however, the equivalent Schönflies symbol for the 32 crystallographic point groups are given in Table 1.3.

Table 1.3 The crystallographic point groups

System	Point group symbols: International	Point group symbols: Shönflies	Number of poles in the general form	Diffraction symmetry
Triclinic	1	C_1	1	$\bar{1}$
	$\bar{1}$	S_2 (C_2)	2	
	2	C_2	2	
Monoclinic	m	C_v	2	$2/m$
	$2/m$	C_{2h}	4	
	mm	C_{2v}	4	
Orthorhombic	222	D_2 (V)	4	mmm
	mmm	D_{2h} (V_h)	8	
	3	C_3	3	$\bar{3}$
	$\bar{3}$	S_6 (C_{3i})	6	
Trigonal	$3m$	C_{3v}	6	
	32	D_3	6	$\bar{3}m$
	$\bar{3}m$	D_{3d}	12	
	4	C_4	4	
	$\bar{4}$	S_4	4	$4/m$
	$4/m$	C_{4h}	8	
Tetragonal	$4mm$	C_{4v}	8	
	$\bar{4}2m$	D_{2d}	8	$4/mmm$
	42	D_4	8	
	$4/mmm$	D_{4h}	16	
	6	C_6	6	
	$\bar{6}$	C_{3h}	6	$6/m$
	$6/m$	C_{6h}	12	
Hexagonal	$6mm$	C_{6v}	12	
	$\bar{6}m2$	D_{3h}	12	$6/mmm$
	62	D_6	12	
	$6/mmm$	D_{6h}	24	
	23	T	12	$m3$
	$m3$	T_h	24	
Cubic	$\bar{4}3m$	T_d	24	
	43	O	24	$m3m$
	$m3m$	O_h	48	

In developing the crystallographic point groups it is convenient to start with those containing a single n-fold axis of symmetry. It is easy to show that the allowed values of n cannot exceed 6: a proof is based on the existence of a minimum length $\boldsymbol{R}$ for a translation vector perpendicular to the rotation axis. Operating on $\boldsymbol{R}$ with the rotation axis yields another allowed vector $\boldsymbol{R}'$, of the same magnitude as $\boldsymbol{R}$ but rotated through $2\pi/n$ (Fig. 1.9a). Since $\boldsymbol{R} - \boldsymbol{R}'$ must also be an allowed translation we can write down that its length must be greater than or equal to $|\boldsymbol{R}|$, the smallest translation, i.e., $2|\boldsymbol{R}| \sin(\pi/n) \geqslant \boldsymbol{R}$, hence $n \leqslant 6$.

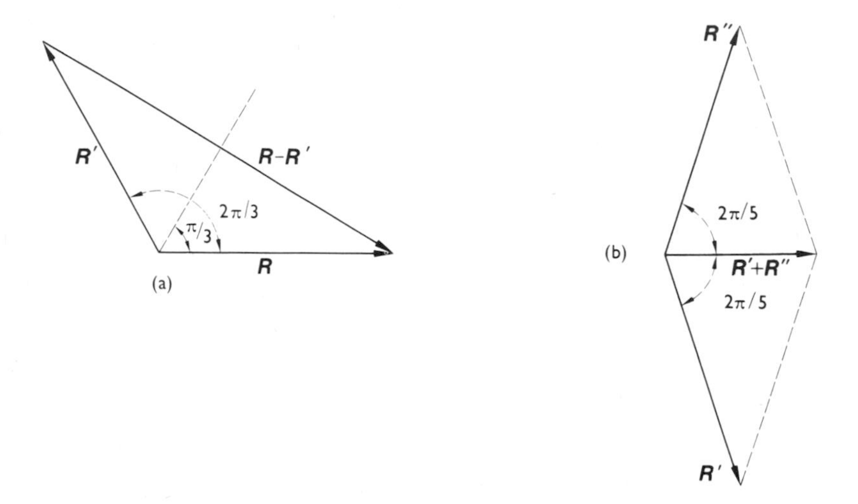

Figure 1.9 The generation of translation vectors perpendicular to (a) a three-fold and (b) a five-fold rotation axis. In (a) $|R - R'| > |R|$, $|R + R'| = |R|$ but in (b) $|R' + R''| < |R|$ and the axis is not consistent with translational symmetry.

A fivefold axis is also excluded since it results in $|\boldsymbol{R}' + \boldsymbol{R}''| < \boldsymbol{R}$ for the rotations shown in Fig. 1.9b. Thus the requirements of translational symmetry restricts rotation axes in crystals to $n = 1, 2, 3, 4$, or 6.

We can now derive stereograms showing the symmetry elements of each crystallographic point group and the group of poles which result from the action of these symmetry elements on a general point. Point group $\bar{3}m$ is chosen to illustrate the method.

Figure 1.10 shows stereograms containing (i) the symmetry elements $\bar{3}m$, (ii) the 12 poles produced by operating these symmetry elements on a general pole and (iii) the complete set of symmetry elements present in the point group which has been obtained by inspection of (ii).

All 32 point groups are illustrated in Fig. 1.11, mirror planes are shown as solid lines and rotation axes of order n by solid n-sided symbols. Open n-sided symbols are used for *inversion* axes, that is rotation by $2\pi/n$

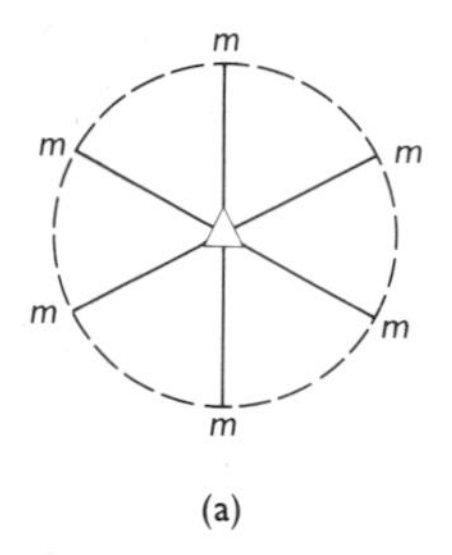

(a)

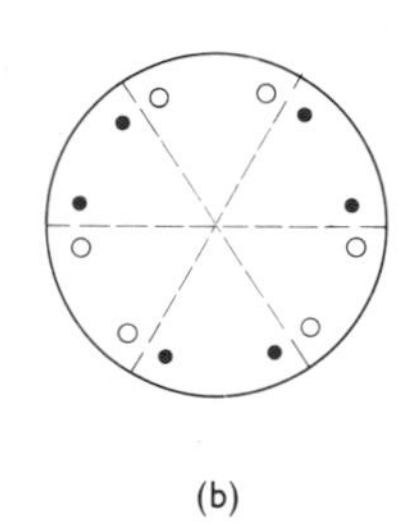

(b)

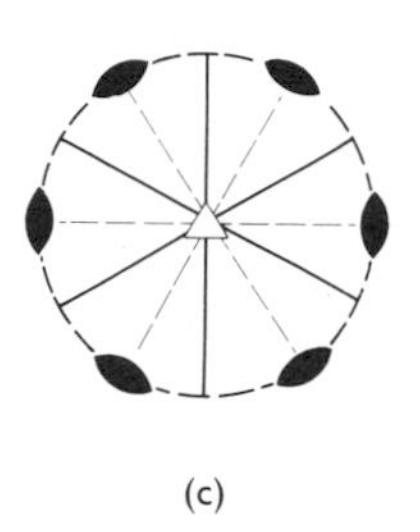

(c)

Figure 1.10 Stereograms of (a) the symmetry elements $\bar{3}m$ (b) $\bar{3}m$ operating on a general pole and (c) all the symmetry elements included by inspecting (b).

followed by inversion through a centre. It should be noted that the same point group may result from different sets of initial symmetry operators, e.g. $3/m = \bar{6}$ and $3/mm = \bar{6}m$: all this means is that the same *total* group of symmetry elements is implied by both groups of symbols. In the symbol $\bar{4}2m$, the order of the subsidiary symbols is important, the first subsidiary denoting the symmetry element chosen to lie along the X axis and the second a symmetry element which occupies a direction inclined to the X axis and not related to it by the principal axis.

The point group symbols in both notations are given in Table 1.3, together with the number of poles in the *general form. Special forms* are groups of poles produced by the point group operating on a pole which bears some special relationship to one or more of its symmetry elements such that the total number of poles is reduced. The cube is a special form in all the cubic point groups.

The most symmetrical point group in each of the crystal systems (the last entry in each block of Table 1.3) is known as the *holosymmetric* point group. The arrangement of other lattice points around any one always shows the holosymmetric symmetry of the system so that the conventionally chosen cell also has this symmetry. The symmetry of the atomic arrangement within such a cell may have less than the holosymmetric symmetry and this lower symmetry may be reflected in the external faces of the crystal. For example, the regular tetrahedron belongs to the cubic system because it has the necessary four triads though it lacks the full symmetry shown by a cube.

1.6 Laue symmetry

The crystal classes may be grouped according to the symmetry they exhibit in their diffraction effects. *Friedel's law* states that, in general, every crystal diffracts as if a centre of symmetry were present. (Exceptions

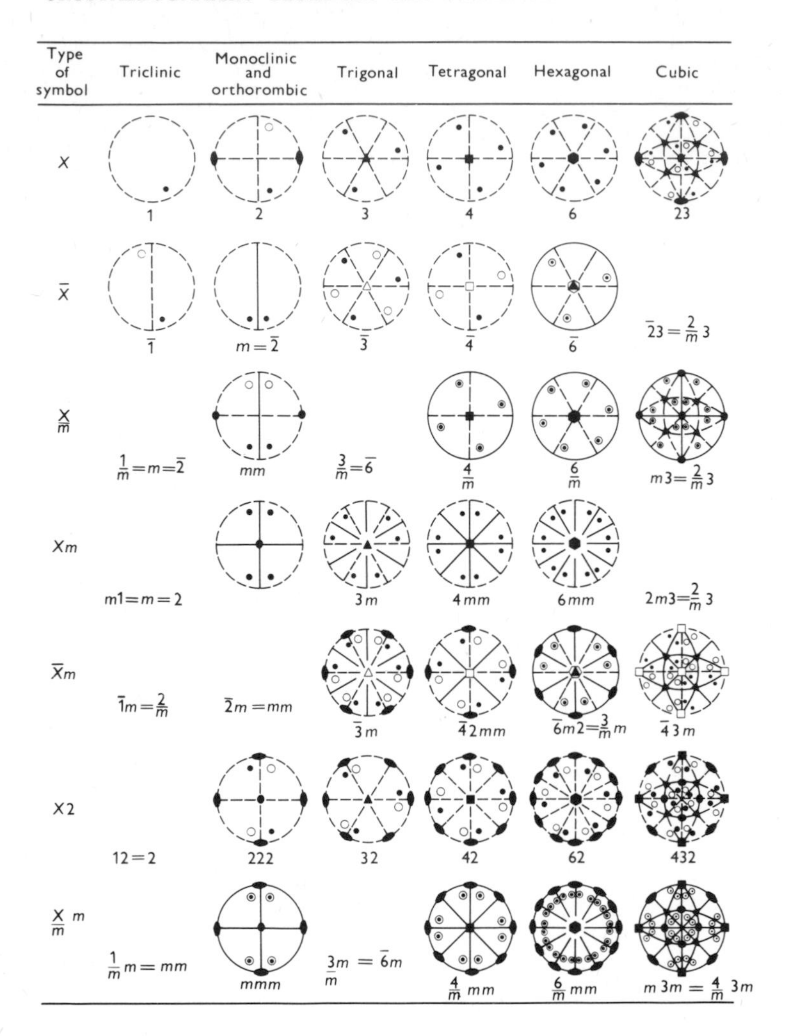

Figure 1.11 Stereograms showing the symmetry elements and the poles of a general form in each of the 32 crystallographic point groups. Mirror planes are indicated by full lines.

to this law occur if the incident wave has an energy close to that of a strong transition between two energy states in the scattering system (see §2.4 and §3.11).) The last column of Table 1.3 lists the 11 Laue symmetry groups which describe the diffraction symmetry exhibited by the 32 crystal classes.

1.7 Non-crystallographic point groups

In the previous two sections we have laid emphasis on the point groups which describe the symmetry of crystals. Many crystal structures exist in which groups of atoms are arranged with a point group whose symmetry is inconsistent with the requirements of translation which are essential for a crystallographic point group. A commonly found example of this type is the occurrence of the point group 532, which describes the 15 twofold, 10 threefold and six fivefold axes of the regular icosahedron. Many intermetallic phases exhibit local areas in which the 12 neighbours to a central atom lie at the corners of an icosahedron and the same figure is prominent in a number of virus particles such as tomato bushy stunt virus (Caspar, 1956) and turnip yellow mosaic virus (Klug *et al.*, 1957). Figure 1.12 shows the way the icosahedral $V{-}Al_{12}$ groups are arranged to conform with an overall cubic unit cell in the intermetallic compound VAl_{10} (Brown, 1957).

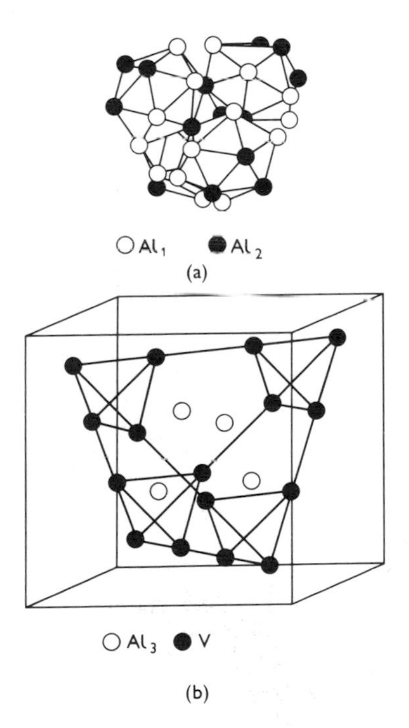

Figure 1.12 Icosahedral groups of aluminium atoms about the vanadium atoms in the structure of VAl_{10}. (a) The vanadium atoms are at the centres of the icosahedra, which are drawn as though solid. (b) The packing within the unit cell: all Al atoms in contact with V are omitted.

1.8 Space groups

We must now consider the symmetry elements required to describe a three-dimensional array of atoms within a crystal. In a point group all the symmetry elements pass through a point. To describe the regularity of a crystal structure, symmetry elements must be arrayed throughout the lattice and in addition to the pure rotation and reflection operators described previously, elements may occur in which their operations are combined with translations. These new elements are called *screw axes* and *glide planes* respectively and the translations which they introduce are simple sub-multiples of lattice vectors. The complete scaffold of elements is termed a *space group* and there are 230 such groups which can describe the symmetry of crystals. Each space group belongs to that specific point group which has the same symmetry elements in the same directions ignoring the translations associated with screw axes and glide planes. The space groups are fully described in the *International Tables for Crystallography*, Volume I, (1952): we shall consider one example, *Pmna*, belonging to the orthorhombic point group *mmm*.

In the space group *Pmna* the first symbol tells us that the Bravais lattice is primitive (see § 1.3). The remaining symbols refer to the three mirror planes in the parent point group; in this space group the one perpendicular to the x axis is a true mirror plane, the one perpendicular to the y axis is an n-glide plane, i.e. one which reflects accompanied by a translation of $a/2$ and $c/2$ and the one perpendicular to the z axis is an a-glide plane, i.e. one which translates by $a/2$ as well as reflecting. Figure 1.13 shows (a) the disposition of these symmetry elements within the unit cell, (b) the positions they generate when a general point is introduced and (c) the additional symmetry elements which are implied in the original description.

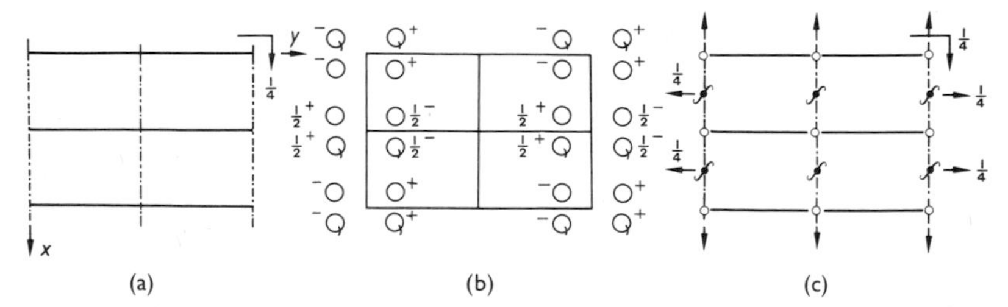

Figure 1.13 The space group *Pmna*; (a) includes only the symmetry elements *m, n* and *a*, (b) shows the positions they generate when a single representative point is introduced and (c) illustrates the complete group of symmetry elements obtained by inspection of (b).

and which can be seen by inspection of (b). It should be noted that the arrangement of equivalent positions is centrosymmetric as is the parent point group, and the number of general positions is the same as the number of poles (8) in the general form of *mmm*. If the cell had been *C*-, *I*- or *F*-centred, however, the number of general positions would have been increased by 2, 2 and 4 times respectively.

1.9 Magnetic symmetry

In Sections 1.5–1.8 we have been considering the symmetry elements which describe relationships between scalar quantities. Many crystals, however, exhibit the phenomena of ferromagnetism and antiferromagnetism in which certain atoms in the solid possess net magnetic moments which have fixed orientations relative to the crystal lattice. The point groups and space groups that we have used previously are inadequate to describe the symmetry relationships which exist between magnetic moments since, in addition to rotation, reflection and translation, we must introduce the concept of *vector reversal.*

In considering the effect of symmetry elements on the magnetic moment, it must be recalled that a magnetic moment is equivalent to a current loop: the moment vector is perpendicular to the current loop and its direction is given by the right-handed screw convention. The operation of reflection in a plane containing the current loop does not reverse the moment direction, whereas a mirror plane perpendicular to the loop, containing the moment direction, reverses the direction of current in the loop and hence the moment. In both cases the result of the reflection operation on magnetic moment, which is an *axial vector,* is the reverse of that on a position vector. This is true for all operations involving mirror inversion.

For any magnetic crystal, each element of its space group should correspond to one of two elements in the magnetic space group. In one of these the operation of the element on the moment is that to be expected for an axial vector and in the other it is accompanied by the time reversal operator. The action of the time reversal operator is to reverse the direction of current in the loop and hence to introduce a moment reversal *in addition to* any implied by the parent element.

We have now completed our discussion of the symmetry properties of crystalline solids, and the remaining sections of this chapter will be devoted to obtaining some geometrical properties of lattices which are of general use and which will be needed in later chapters. The most popularly recognised feature of crystals is their well-formed faces and it may be asked how these faces relate to the internal structure and the crystal lattice. Crystals form faces whose planes have a high density of atoms. Such planes pass through many unit cells of the crystal and a continuous high density will be maintained only if the lattice vectors within the plane are short.

1.10 Miller indices

If the unit cell of a crystal is defined by the vectors $\boldsymbol{a}$, $\boldsymbol{b}$ and $\boldsymbol{c}$, then any crystal plane which intercepts the axes in lengths proportional to a/h, b/k, c/l respectively is denoted by its Miller indices (hkl) [Fig. 1.14]. The Miller indices of the plane making negative intercepts a/h, b/k, c/l are conventionally written $(\bar{h}\bar{k}\bar{l})$. The equations of the normal $\boldsymbol{p}$ to the face can be written down in terms of its direction cosines

$$(a/h)\cos(\widehat{\mathrm{pa}}) = (b/k)\cos(\widehat{\mathrm{pb}}) = (c/l)\cos(\widehat{\mathrm{pc}}) \tag{1.2}$$

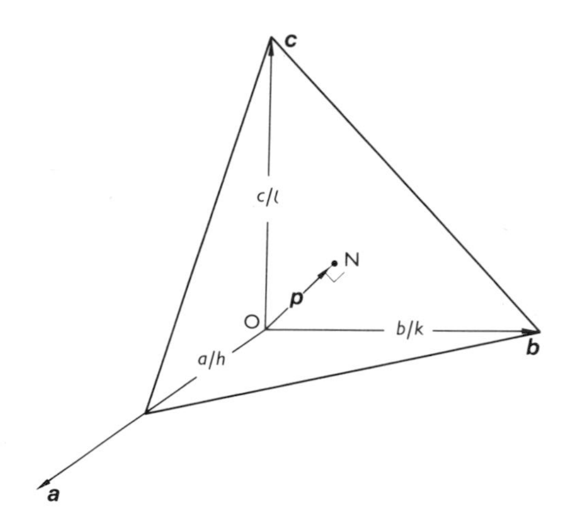

Figure 1.14 The Miller indices of the plane with intercepts $a/2$, b and c is (211).

The requirement for short lattice vectors to occur in a plane is that the Miller indices should be small integers, which is the *law of rational indices.* It is sometimes convenient to use four axes in the trigonal and hexagonal systems, since the rotation axes imply the existence of three equivalent axes at 120° intervals in the plane perpendicular to the rotation axis ($\boldsymbol{c}$). This is the Miller–Bravais system of notation, the third axis being denoted u with corresponding index i. The first three face indices are not independent since $h + k + i = 0$: the third index is sometimes suppressed for this reason, the face $(21\bar{3}1)$ being denoted (21.1).

A Miller index (*hkl*) will still define the same face normal *direction* when it is divided through by any common factor and it is conventional to use the reduced index when describing the external morphology of a crystal. However, the interplanar spacing $\boldsymbol{p}$ (Fig. 1.14) will be changed by this process, being multiplied by the common factor. We shall see in Section 3.5 that the interplanar spacing is particularly important in describing the diffraction from crystals and for this purpose any common factors must be retained.

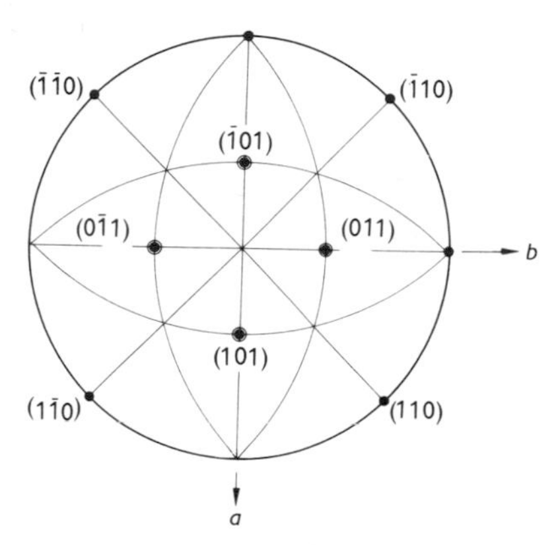

Figure 1.15 The Miller indices corresponding to faces of a rhombic dodecahedron. Only faces in the upper hemisphere have been indexed, the indices corresponding to open circles will have negative l indices and would be denoted $(10\bar{1})$ $(\bar{1}0\bar{1})$ $(01\bar{1})$ and $(0\bar{1}\bar{1})$.

The interplanar spacing d corresponding to a plane (*hkl*) in a triclinic crystal is given by

$$\frac{1}{d} = \sqrt{\frac{\frac{h}{a}\begin{vmatrix} \frac{h}{a} & \cos\gamma & \cos\beta \\ \frac{h}{b} & 1 & \cos\alpha \\ \frac{l}{c} & \cos\alpha & 1 \end{vmatrix} + \frac{k}{b}\begin{vmatrix} 1 & \frac{h}{a} & \cos\beta \\ \cos\gamma & \frac{k}{b} & \cos\alpha \\ \cos\beta & \frac{l}{c} & 1 \end{vmatrix} + \frac{l}{c}\begin{vmatrix} 1 & \cos\gamma & \frac{h}{a} \\ \cos\gamma & 1 & \frac{k}{b} \\ \cos\beta & \cos\alpha & \frac{l}{c} \end{vmatrix}}{\Delta}} \qquad (1.3)$$

where Δ is the determinant

$$\begin{vmatrix} 1 & \cos\gamma & \cos\beta \\ \cos\gamma & 1 & \cos\alpha \\ \cos\beta & \cos\alpha & 1 \end{vmatrix}$$

For a cubic crystal this reduces to

$$d = a/\sqrt{h^2 + k^2 + l^2} \tag{1.4}$$

A simpler expression can be given for the triclinic case in terms of the reciprocal lattice constants (see §3.4 and §3.5).

1.11 Zone axis symbols and the zone law

The *zone axis symbol*, written $[UVW]$, defines the direction of the zone axis as that from the origin to the point $(U\mathbf{a} + V\mathbf{b} + W\mathbf{c})$. Any two faces $(h_1 k_1 l_1)$, $(h_2 k_2 l_2)$ define a zone axis which is parallel to their line of intersection and perpendicular to the plane containing the face normals. Since the equations of planes through the origin and parallel to the faces are

$$h_1 \frac{x}{a} + k_1 \frac{y}{b} + l_1 \frac{z}{c} = 0 \quad \text{and} \quad h_2 \frac{x}{a} + k_2 \frac{y}{b} + l_2 \frac{z}{c} = 0$$

the equations of the zone axis are

$$\frac{x}{a(k_1 l_2 - l_1 k_2)} = \frac{y}{b(l_1 h_2 - h_1 l_2)} = \frac{z}{c(h_1 k_2 - k_1 h_2)} \tag{1.5}$$

and

$$\begin{aligned} U &= k_1 l_2 - l_1 k_2 \\ V &= l_1 h_2 - h_1 l_2 \\ W &= h_1 k_2 - k_1 h_2 \end{aligned} \tag{1.6}$$

Multiplication of these three equations by h_1, k_1 and l_1 respectively, followed by summation gives an immediate proof of the *Weiss zone law*

$$hU + kV + lW = 0$$

If $(h_1 k_1 l_1)$ and $(h_2 k_2 l_2)$ lie in the same zone $[UVW]$ a simple device known as 'cross-multiplication' is helpful in evaluating UVW. Each Miller index is written down twice and the appropriate products are obtained by multiplying the pairs of numbers indicated by the arrows, subtracting the product for which the arrow points from right to left.

$$\begin{array}{c|cccc|c} h_1 & k_1 & l_1 & h_1 & k_1 & l_1 \\ h_2 & k_2 & l_2 & h_2 & k_2 & l_2 \end{array}$$

There remains an ambiguity in the sign of $[UVW]$ since writing $h_2 k_2 l_2$ first results in $[\bar{U}\bar{V}\bar{W}]$. Both symbols define the same line, the direction derived being the one which completes a right-handed set with the two face normals.

We may determine by way of illustration the zone axis defined by the faces (011) and (123) as $[11\bar{1}]$

$$\begin{array}{c|cccc|c} 0 & 1 & 1 & 0 & 1 & 1 \\ 1 & 2 & 3 & 1 & 2 & 3 \\ \hline \multicolumn{6}{c}{1 \quad 1 \quad \bar{1}} \end{array}$$

A similar relationship exists for the determination of the indices of a face which lies in two zones: if the zones are [111] and $[\bar{1}01]$ the common face is

$$\begin{array}{c|cccc|c} 1 & 1 & 1 & 1 & 1 & 1 \\ \bar{1} & 0 & 1 & \bar{1} & 0 & 1 \\ \hline \multicolumn{6}{c}{(1 \quad \bar{2} \quad 1)} \end{array}$$

Care must be taken in interpreting zone axis symbols for hexagonal axes using Miller–Bravais notation: it is usual to omit the non-independent index corresponding to the u axis from all calculations and the resultant co-ordinates are referred to the three remaining axes and written as $[UV\dagger W]$. The figure $-(U + V)$ must *not* be inserted in such a zone symbol since in general $U\hat{\boldsymbol{x}} + V\hat{\boldsymbol{y}} \neq -W\hat{\boldsymbol{u}}$, where $\hat{\boldsymbol{x}}$ etc. denote unit vectors parallel to the three hexagonally equivalent axes. When, however, a calculation leads to a face index $(hk.l)$, the third figure can be re-inserted using the relationship $h + k + i = 0$. For example, the zones containing $(20\bar{2}1)$, $(01\bar{1}0)$ and $(10\bar{1}0)$, $(01\bar{1}1)$ can be found to have zone axis symbols

$$\begin{array}{c|cccc|c} 2 & 0 & 1 & 2 & 0 & 1 \\ 0 & 1 & 0 & 0 & 1 & 0 \\ \hline \multicolumn{6}{c}{[\bar{1}\ 0\dagger 2]} \end{array} \qquad \begin{array}{c|cccc|c} 1 & 0 & 0 & 1 & 0 & 0 \\ 0 & 1 & 1 & 0 & 1 & 1 \\ \hline \multicolumn{6}{c}{[0\ \bar{1}\dagger 1]} \end{array}$$

The face common to them both is given by

$$\begin{array}{c|cccc|c} 1 & 0 & 2 & 1 & 0 & 2 \\ 0 & \bar{1} & 1 & 0 & \bar{1} & 1 \\ \hline \multicolumn{6}{c}{(2\ \ 1.1)} \end{array}$$

with the full index $(21\bar{3}1)$.

In metallurgical texts a second convention for hexagonal zone axis symbols is frequently adopted in which four symbols are present and the sum of the first three is zero. The zone axis direction $[UVSW]$ given by the vector from the origin to the point $U\boldsymbol{a} + V\boldsymbol{b} + S\boldsymbol{e} + W\boldsymbol{c}$, where $a = b = e$ and their directions are parallel to the three equivalent hexagonally related axes. Using this convention the zone law becomes $hU + kV + iS + lW = 0$.

A convention exists for the use of brackets around three integer symbols such as the Miller or zone indices:

(hkl)	denotes the Miller indices of a plane.
$\{hkl\}$	implies all those planes (hkl) related by the point group symmetry, namely all faces of the *form* (see §1.5).
$[UVW]$	is a zone axis symbol.
$\langle UVW \rangle$	implies all the zone axes related by the point group symmetry.

1.12 The angles between zones and faces

The general equations for the angles between two zones, a face and a zone and between two faces are complicated, but very simplified forms exist for the cubic system, where the zone axis $[UVW]$ defines the same direction as the face normal (hkl), where $U = h$ and $V = k$ and $W = l$. For the cubic system, the angle ϕ between two faces $(h_1 k_1 l_1)$, $(h_2 k_2 l_2)$ is given by

$$\cos \phi = \frac{h_1 h_2 + k_1 k_2 + l_1 l_2}{\sqrt{h_1^2 + k_1^2 + l_1^2}\,\sqrt{h_2^2 + k_2^2 + l_2^2}} \tag{1.7}$$

The angle between (111) and (110) in the cubic system would therefore be given by

$$\cos (111) \wedge (110) = \frac{1+1}{\sqrt{3}\sqrt{2}} = \frac{\sqrt{2}}{\sqrt{3}} = 0{\cdot}8163$$

$$(111) \wedge (110) = 35^\circ 16'$$

Tabulations of the angles between a large number of low index faces in the cubic system are given in the *International Tables for Crystallography*, Volume II, (1959) together with the more general equations applicable to crystal systems of lower symmetry. Fixed sets of interplanar angles also occur between normals in the plane perpendicular to the unique axis for the trigonal, hexagonal and tetragonal systems.

2

The Production and Properties of X-rays, Neutrons and Electrons

2.1 Introduction

In the first chapter of this book we have tried to show how the external symmetries of crystals are related to their internal regularities, and to provide a framework within which the essential geometry of crystal structure can be described. The external geometry and symmetry of crystals is easily recognised in well-formed specimens and the measurement and description of these external features was for a long time the main concern of crystallographers. However since 1912, the year in which the first X-ray diffraction pattern was observed, a complete revolution has taken place in the science of crystallography. It has been possible to deduce in full detail the internal structures of countless solids, many of which from their external appearance alone would never have been recognised as crystals. The foundation upon which this revolution has been based is the phenomenon of crystal diffraction, primarily X-ray diffraction but strongly supplemented in recent years by electron and neutron diffraction. It is to a study of these methods and to the results obtained by them that the rest of this book is devoted and it is appropriate, therefore, at this point to give a brief account of the production and some of the essential properties of these radiations.

2.2 Production of X-rays

X-Rays are emitted when a beam of electrons strikes a surface. They have been shown to form part of the electromagnetic spectrum with wavelength of the order of 1 Å. The precise spectrum of the X-rays emitted depends both on the material of the surface and on the energy of the electron beam. A commonly used type of X-ray diffraction tube is shown schematically in Fig. 2.1.

Electrons are emitted from the hot filament which is a coil of tungsten wire held at a negative potential with respect to the metal anode which is earthed. Electrons emitted from the filament are focused by the space-charge distribution and by the geometrical form of the cathode; they are accelerated by the applied potential and strike the anode over an area about 1 x 0·1 cm in size. X-Rays are emitted from this area and some of them pass out of the tube through beryllium windows in each of its four sides.

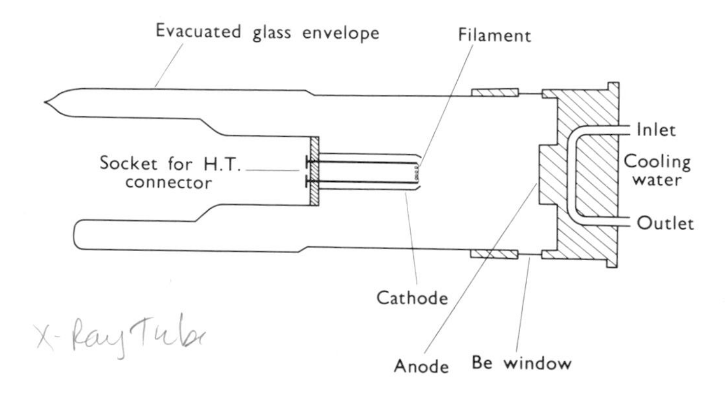

Figure 2.1 Schematic diagram of a sealed-off X-ray tube. The filament and cathode assembly is held at a high negative potential with respect to the anode which is earthed. X-rays emitted from the surface of the anode emerge through the beryllium windows.

The windows are positioned so as to let through X-rays emitted at angles around 6° to the surface of the anode, thus the projected emitting area observed through one pair of windows is about 0·1 x 0·1 cm (square focus) and that through the other two windows 0·01 x 1·0 cm (line focus). The anode must be made of a highly conducting material so that the heat generated can rapidly be transferred to the cooling water. If it is necessary to use a target which is a poor conductor, such as molybdenum, it is usually prepared as a thin layer plated or sprayed on to the surface of a good conductor such as copper. A glass envelope, which carries the beryllium windows, encloses both anode and cathode. It is highly evacuated during the manufacturing process and then sealed.

2.3 Spectrum of X-rays emitted

The spectrum of X-rays obtained by bombarding a copper target with 35 keV electrons is illustrated in Fig. 2.2.

It can be seen to consist of a broad band of continuous (white) radiation together with a number of discrete emission lines. Both parts of the spectrum can be understood in terms of quantum theory, the continuous radiation results from inelastic collisions between the exciting electrons and

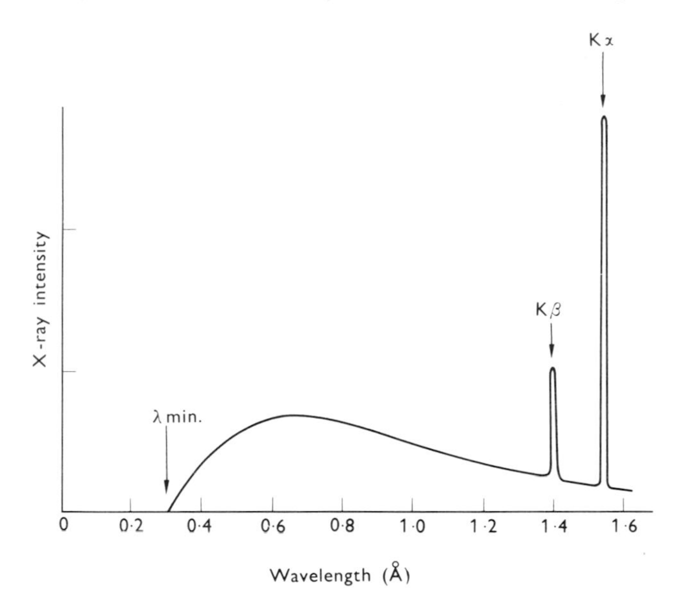

Figure 2.2 Spectrum of X-rays emitted by a Cu target at 35 keV excitation.

those of the target material; energy lost in the collisions is emitted as X-radiation. The white spectrum has a well-defined minimum wavelength (maximum energy) which corresponds to a collision in which an electron loses the whole of its energy in a single process. Thus

$$\lambda_{\text{min.}} = \frac{hc}{eV_0} \quad \text{or} \quad \lambda_{\text{min.}} = 12{\cdot}34/V_0 \tag{2.1}$$

where λ is in Å and V_0 in kV.

It is found in practice that the maximum intensity of the white spectrum occurs at a wavelength approximately twice $\lambda_{\text{min.}}$. The characteristic line spectrum of the target is emitted during the rearrangement of the orbital electrons of the target atoms consequent on the ejection of one or more of their inner electrons by the bombarding electrons. The K emission lines correspond to radiation emitted when electrons fall from the outer shells to fill holes in the K shell; the Kα line corresponds to an L–K transition and the $K\beta_1$ line to an M to K transition. Similarly, the L and M lines correspond to transitions into the L and M shells respectively, the α lines corresponding to transitions with a change of unity in the principal quantum number and the strongest β and γ lines to changes of 2 and 3 respectively. The selection rules for these transitions are such that the $K\alpha$ line is a close doublet with the stronger component $K\alpha_1$ having twice the intensity of the weaker $K\alpha_2$ component. As the atomic number of the target element increases, the energy difference between successive shells increases and hence the wavelength of the characteristic K lines is shortened. Table 2.1 gives a list of the wavelengths of the $K\alpha$ and $K\beta$ lines of some commonly used target materials.

Table 2.1 Data for some common target materials†

Target element	Wavelength (Å)					β-Filter	
	$K\alpha_2$	$K\alpha_1$	Mean $K\alpha$	$K\beta_1$	K absorption edge	Element	Thickness $\times 10^{-3}$ (cm)
Cr	2·2935	2·2896	2·2909	2·0848	2·070	V	1·7
Mn	2·1057	2·1017	2·1031	1·9101	1·896	Cr	1·7
Fe	1·9399	1·9360	1·9370	1·7565	1·743	Mn	1·8
Co	1·7928	1·7889	1·7902	1·6208	1·608	Fe	2·0
Ni	1·6617	1·6578	1·6591	1·5001	1·488	Co	2·0
Cu	1·5443	1·5405	1·5418	1·3922	1·380	Ni	2·2
Mo	0·7135	0·7093	0·7107	0·6322	0·620	Zr	11·3
Rh	0·6176	0·6132	0·6147	0·5456	0·533	Ru	7·3
Ag	0·5638	0·5594	0·5608	0·4970	0·486	Rh	7·7

† The filter thickness given is that required to reduce the intensity of the $K\beta$ line relative to the $K\alpha$ by a factor of 100.

2.4 Absorption of X-rays

When an X-ray beam passes through matter its intensity is attenuated by an amount which depends upon the thickness and density of the material and also on the wavelength of the radiation. For any material and wavelength it is possible to define a linear absorption coefficient μ such that the ratio of the intensity (I) transmitted normally through a thickness t, to the incident intensity (I_0) is given by

$$I/I_0 = e^{-\mu t} \tag{2.2}$$

The absorption of an element can be described by a mass absorption coefficient μ_m such that

$$\mu_m = \mu/\rho \tag{2.3}$$

where ρ is the density.

To a good approximation the absorption of a compound or mixture depends only on the elements present and their proportions and not on the state of combination. The linear absorption coefficient of a multi-component material can therefore be expressed as

$$\mu = \rho \sum_i p_i \mu_{mi} \tag{2.4}$$

where μ_{mi} is the mass absorption coefficient of the ith element in the material and p_i the fraction by weight of that element, the sum being taken over all elements present. There is one further absorption coefficient which is often tabulated; it is the gram-atomic coefficient μ_g which is given by

$$\mu_g = \mu_m W \tag{2.5}$$

where W is the atomic weight of the element. The gram-atomic absorption coefficient is useful in calculating the absorption in chemical compounds whose formulae are known. In such a case the linear absorption coefficient

$$\mu = \rho \sum_i n_i \mu_{gi} / \sum n_i W_i \tag{2.6}$$

where n_i is the number of atoms of element i in one formula unit, μ_{gi} its gram-atomic absorption coefficient and W_i its atomic weight.

The wavelength dependence of absorption coefficients is illustrated in Fig. 2.3 which shows the mass absorption coefficients of nickel and of barium as a function of wavelength. It can be seen that there is a general tendency for the absorption coefficient to increase with increasing wavelength. Superimposed on this increase are the effects of a number of '*absorption edges*'; at wavelengths just below the edges there is a steep increase in the absorption and at the edge itself there is a nearly vertical fall in the coefficients to a value which may be a factor 10 smaller than the value just before the edge. These absorption edges occur at X-ray energies near which there is a resonant interaction between the incident radiation and one of the electronic levels in the atoms of the absorber, thus each absorption edge can be linked with a group of X-ray emission lines of the absorbing element. The K absorption edge corresponds to the resonant ejection of a K shell electron into the continuum; it is therefore at a wavelength close to, but rather shorter than, the K emission lines. It is perhaps worth mentioning that when the wavelength of the radiation is near to an absorption edge there is a significant change in the scattering power of the atom and hence of its atomic scattering factor (see §3.6). This phenomenon, usually referred to as *anomalous scattering*, is due to the resonant interaction and leads to a change in phase of the scattered radiation at wavelengths just below the edge where the absorption is high. At such wavelengths the atomic scattering factor becomes a complex quantity and it is possible to observe departures from Friedel's law (see §1.6 and §3.11). At wavelengths below the absorption edge there is of course excitation of fluorescent radiation as the electrons relax into the holes left vacant by the

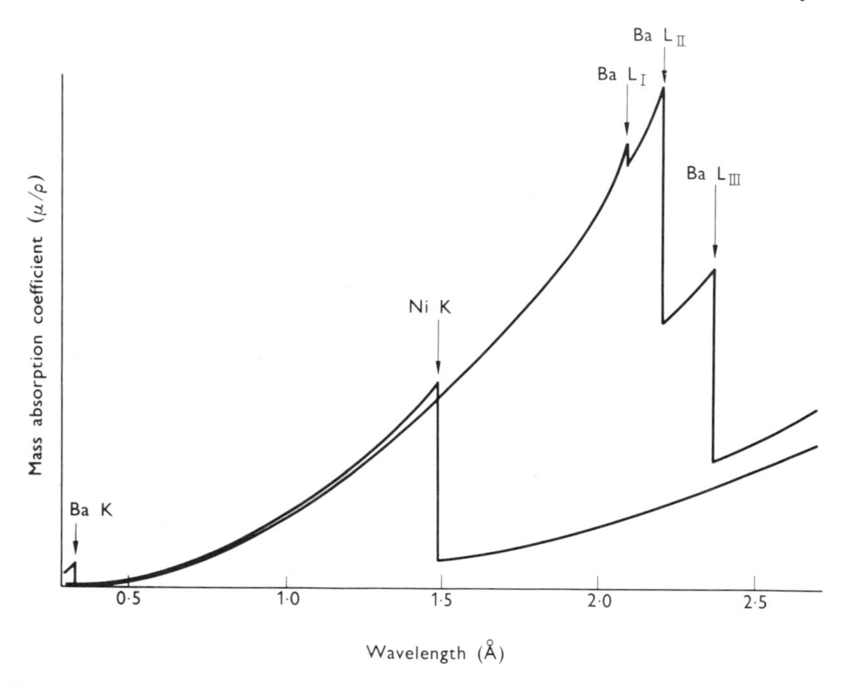

Figure 2.3 The wavelength dependence of the X-ray mass absorption coefficients of nickel and barium.

resonant excitation. This radiation often gives an unwanted high background level to X-ray photographs; it is therefore necessary to consider the possibility of fluorescence when selecting a suitable radiation for use in a particular problem. For a sample containing copper, Cu K radiation could be used because its energy is just insufficient to excite fluorescent K radiation from the copper of the sample; however, if the sample contained nickel, fluoresence would occur with the copper Kβ radiation and to avoid a high background it would be necessary either to filter out the Kβ radiation or to use a target material giving a longer wavelength.

One use of the K absorption edges is in enabling the Kβ lines to be attenuated with respect to the Kα, so that an approximately monochromatic X-ray beam can be obtained for diffraction purposes. The appropriate filter material is an element whose atomic number is less than that of the target material by one or perhaps two units. The K absorption edge of the filter should lie between the Kα and Kβ lines of the target and as close to the Kβ line as possible. Under these conditions the absorption coefficient of the filter for Kβ radiation will be much greater than that for Kα and hence, by using an appropriate thickness of filter, the transmitted radiation can be made to consist mainly of the Kα line. The appropriate β filters for the commonly used radiations and the thicknesses required to attenuate the β radiation relative to the α by a factor of 100 are listed in Table 2.1.

2.5 Production of electron beams for diffraction purposes

Electron beams, whether for diffraction or for other purposes, are almost invariably produced by thermionic emission from a heated cathode. After emission the electrons are accelerated to the required energy by passing through an electric field and are focused by electrostatic and magnetic lenses into a narrow beam. The path of the beam must be in a high vacuum both to avoid electrical breakdown under the applied potentials and to prevent absorption and scattering of the electrons themselves by gas molecules. The electron gun and incident beam focusing system are very much the same as those of an electron microscope and, in fact, many electron microscopes are designed so that they can also be used for obtaining diffraction photographs. Such systems, however, often have the disadvantage of rather small apertures so that only spectra diffracted at relatively low angles can be observed. This difficulty is overcome in apparatus designed specifically for electron diffraction, and Fig. 2.4 is a schematic drawing of the electron-diffraction apparatus used by Vainstein and Pinsker (1958).

The electron gun can be used with accelerating potentials up to 100 kV and consists of a heated filament held at a positive potential with respect to the surrounding cylindrical hood which has an exit hole 0·1 mm in diameter. The anode section contains a diaphragm through which the beam passes at the end of its accelerated path. The magnetic lens is used to focus the beam in the plane of the fluorescent screen when diffraction images are being observed. An alternative position for focusing is just in front of the specimen, in which case a shadow image with a magnification of about 50 times is obtained on the screen. The position and orientation of the specimen at A can be adjusted from outside the vacuum envelope using rods actuated through bellows couplings. An alternative specimen position which gives a larger range of scattering angles is provided at B.

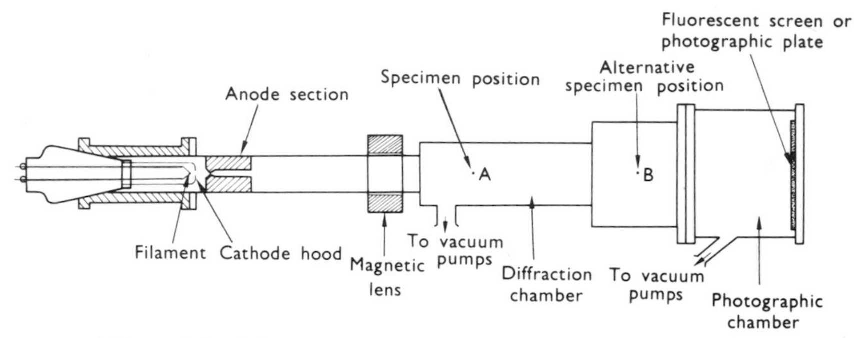

Figure 2.4 Schematic diagram of an electron diffraction apparatus.

Electrons with the energies used in diffraction are very strongly absorbed in matter. Specimens which are to be used in transmission must therefore be in the form of thin films a few hundreds of Ångstrom units thick and it must be remembered, when reflection photographs are taken of thicker specimens, that only the surface layers of the specimen contribute to the diffraction pattern.

The small penetration of low-energy electrons into a crystal is exploited in the technique commonly known as LEED (low-energy electron diffraction). This technique is used to study the structure of surfaces using electrons in the energy range 10–500 eV (150 eV $\simeq$ 1 Å). The penetration of such electrons into a solid is limited to between 3 and 10 Å, i.e. from one to three atomic layers; hence the scattering depends very largely on the two-dimensional structure of the solid's surface. Electrons scattered in reflection from the surface of the sample are detected either by a Faraday collector or on a fluorescent screen. A review of the technique is given by Bauer (1969).

2.6 The production of thermal neutron beams

Diffraction of thermal neutrons was first observed in 1936 by Von Halben and Prieswerk and it was recognised, at about this time, that neutron scattering could provide a powerful tool for the investigation of crystalline solids. However the only neutron sources available then were of radioactive origin (and hence of very low flux) so that the potential of the method could not be realised. With the advent of nuclear reactors the position has been entirely changed, since such reactors are essentially neutron generators.

In a simple beam reactor the fuel elements, which may be of natural uranium or of uranium enriched in the ^{235}U isotope, are embedded in a moderator and form the reactor 'core'. In most modern beam reactors the moderator is either water or heavy water. The fast neutrons which are emitted when a ^{235}U nucleus undergoes fission are slowed down by collisions with the atoms of the moderator and are then more likely to cause fission in further ^{235}U nuclei. The reactor will operate if, on average, at least one of the neutrons emitted in a fission causes a further fission. The output power of a reactor is controlled by the insertion of rods made of material (usually cadmium) which has a high capture cross-section for thermal neutrons. The reactor core is surrounded by a neutron reflector, which must be a good scatterer and low absorber of neutrons; these are just the properties required of a good moderator and the two may be made of the same material. The core and reflector must be surrounded by a biological shield of suitable material and thickness to attentuate to a safe level all radiations emitted by the reactor. Neutron beams for experimental purposes pass down tubes which pierce the shield and usually lie in the horizontal plane passing through the mid-point of the core. Figure 2.5 shows a much simplified plan of an experimental beam reactor.

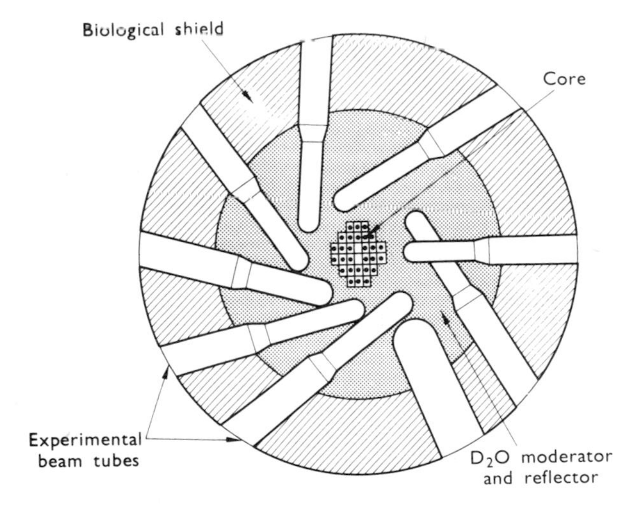

Figure 2.5 Simplified plan of a high flux beam reactor showing the lay-out of horizontal beam tubes.

The distribution of energies of the neutrons emerging from the reactor depends to some extent on the direction of the tubes with respect to the core. Ideally, if the moderator were sufficiently thick the neutrons would be completely thermalised and the energy distribution would be Maxwellian at the temperature of the moderator. In practice this situation is not realised and the beams emerging from radial tubes (those pointing at the core) contain a relatively high proportion of fast and epithermal neutrons and also significant numbers of γ-rays. On the other hand, tubes pointing tangentially to the core give beams containing a significantly higher proportion of thermal neutrons and many fewer γ-rays.

2.7 Monochromatisation of neutron beams

In all cases the spectrum of neutron wavelengths available from the beam tubes is continuous and there is no thermal neutron equivalent of the characteristic X-ray lines. Usually, a necessary preliminary to diffraction experiments is to obtain a 'monochromatic' neutron beam containing only a small range of wavelengths. There are two methods by which such monochromatisation may be carried out, the first of these being by velocity selection using a chopper. The chopper consists of an absorbing disc which has a suitably shaped channel cut through it. The disc is spun rapidly about its axis so that only those neutrons will be transmitted which are incident on the opening as it passes the line of flight and travel with a velocity close to $(2r\omega)/(2\pi - \phi)$, where ω is the angular velocity of rotation and ϕ the angle between the entrance and exit holes. The transmitted beam consists of pulses of neutrons all travelling with approximately the same velocity. The range of velocities (and hence of wavelengths transmitted) depends on the width of the channel through the rotating disc and on its angular velocity. A considerable improvement in wavelength discrimination is obtained by using two or more choppers in series, each carefully phased relative to the others. Such a system can give pulses of neutrons of about 6 μs duration with a wavelength spread of as little as 1%.

The second, and much more commonly used, method for obtaining a monochromatic neutron beam is by using a Bragg reflection (see §3.5) from a large single crystal to select a small range of wavelengths. The crystal may either be used in reflection (cut so as the Bragg planes are parallel to the large surface) or in transmission, with the Bragg planes perpendicular to the surface. The former method is theoretically the more efficient but, because of geometrical convenience, the latter is more often used. With careful selection of a monochromating crystal, up to 60% of the neutrons within the selected wavelength band will be reflected. The most commonly used crystals are beryllium, copper and germanium. The range of wavelengths reflected depends on the degree of collimation of the incident and reflected beams, and on the degree of perfection of the monochromating crystal. Such a monochromator will give a continuous beam of neutrons with wavelength resolution of around 4%. One disadvantage of crystal-monochromatised radiation is that a crystal set to reflect radiation of wavelength λ from the (h, k, l) planes will reflect radiation of wavelength $\lambda/2$ from the $(2h, 2k, 2l)$ planes. For this reason it is usual to work at wavelengths close to one Ångström unit, which is on the short wavelength side of the peak of the neutron spectrum of the reactor; the half-wavelength radiation is then much reduced in intensity because of the sharp fall at short wavelengths of the spectral intensity curve. Another method of overcoming this difficulty is to use the (111) reflection of germanium or silicon because the scattering power of the (222) planes is zero in this instance.

2.8 Absorption of neutrons

Neutrons in general interact rather weakly with matter because of their electrical neutrality. A neutron beam is therefore only slightly attenuated by passing through most solids and consequently neutron absorption coefficients are usually much smaller than the corresponding X-ray absorption coefficients. There are a few notable exceptions provided by elements such as boron, cadmium and gadolinium which have a high capture cross-section for neutrons in the wavelength range around 1 Å. The nuclear reaction involved in neutron capture by each of these nuclei has a different resonant energy; neutrons with energies equal to or greater than this can excite the state corresponding to the combined nucleus and are therefore heavily absorbed. Neutrons with energies just below the resonant values cannot excite this state and are therefore not absorbed, although they are strongly scattered because of the existence of the virtual combined state. This effect is the neutron equivalent of an X-ray absorption edge, a sharp discontinuity in the absorption coefficient, an enhancement of the scattering cross-section at energies just below the edge and the occurrence of an imaginary contribution to scattering (90° phase difference) at energies just above the edge. As in the case of X-rays, the occurrence of an absorption edge may be turned to advantage in the construction of useful filters; thus Pu, which has a resonant capture at 0·52 Å, may be used to discriminate against the $\lambda/2$ component of crystal-monochromatised radiation of around 1 Å wavelength.

3

General Theory of Crystal Diffraction

3.1 Introduction

The measurement of scattering has played a fundamental role in the development of modern physics. Recognition of the wave character of matter requires that all scattering will give rise to diffraction on a scale which will be observable if the size of the scattering object is comparable with the wavelength associated with the scattered radiation. The scattering from any object whose dimensions are much smaller than this wavelength will be independent of angle (isotropic): the scattering from any uniform object whose dimensions are much larger will be governed by the rules of geometric optics. In the intervening range of size, the angular distribution of the scattered amplitude can be described by a form factor and this form factor is determined by the distribution of scattering matter within the object. The reader will already be familiar with the form applicable to slits and discs in optical diffraction: for example, the Airy function shown in Fig. 3.1a. A less obvious example is the electric form factor of the proton for high energy electron (up to 200 GeV i.e. down to 5×10^{-15} cm wavelength) scattering shown in Fig. 3.1 b; the details of this have enabled the charge distribution within the proton (radius $\sim 10^{-13}$ cm) to be deduced (see for example Wilson and Levinger, 1964).

Scattering can be observed from crystals so long as the radiation interacts in some way with the constituent atoms. The diffraction from the crystal as a whole may be considered in two stages: firstly the scattering from a single atom, which is described by its form factor; secondly the diffraction due to the regular arrangement of these atoms within the crystal. It will be clear from the above considerations that the form factor depends upon the nature and the range of the interaction between the radiation and the atom. For example, since the range of interaction between the nucleus and the neutron is of the order of 10^{-13} cm, the form factor for scattering of neutrons with wavelength 1 Å (10^{-8} cm) will be essentially isotropic. Diffraction due to the lattice structure is similar in character to that of a diffraction grating. The three-dimensional nature of the periodicity in a crystal concentrates the diffracted radiation into specific directions. As in the case of the optical grating, the angular separation of the diffraction maxima is proportional to the wavelength and inversely proportional to the periodicity. Indeed, this reciprocal relationship between the characteristic lengths of the object and its scattering pattern is common to all diffraction.

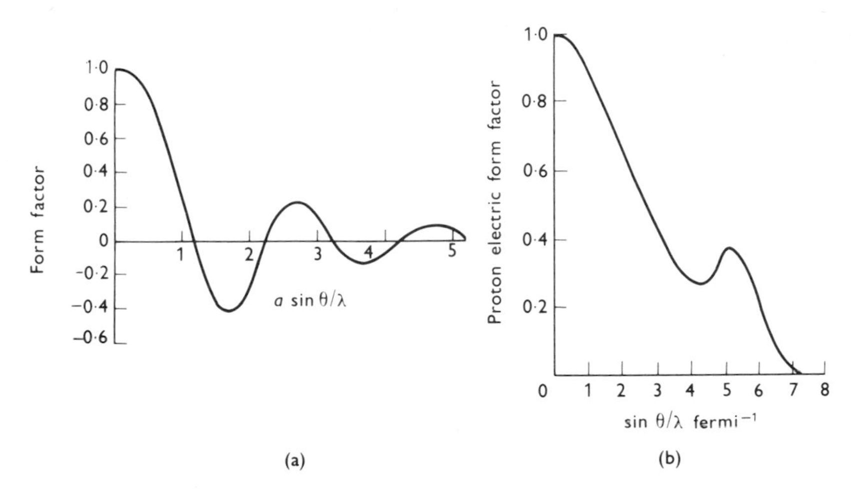

Figure 3.1 (a) Form factor for scattering of light by a disc-shaped opening of radius *a*. (b) Proton electric form factor for high energy electron scattering (data given by Wilson and Levinger, 1964). 1 fermi = 10^{-15} m.

Apart from the precise physical mechanism of the interaction, the conditions which govern the formation of diffraction patterns are very similar for all types of wave disturbance. In this chapter therefore we shall discuss the general features of diffraction by crystals which are common to all types of wave. The physical mechanisms giving rise to electron, X-ray and neutron diffraction and the special features which these introduce form the subject of Chapter 4.

3.2 The general equation for weak scattering

Throughout this section, and most of those that follow, we shall assume that the crystal is a weak scatterer so that the process of scattering does not have a significant effect on the incident wave. This also implies that the interaction between the wave field and the crystal is a small perturbation of the conditions outside the crystal, so that the scattering can be calculated using first-order perturbation theory. The extent to which this approximation is not valid will be considered in Section 4.17. We also assume that the distances between source and crystal, and between crystal and detector, are very long compared with the wavelength so that the conditions are those for Fraunhofer diffraction. This assumption is fully justified since the wavelength is of the order 10^{-8} cm and the other distances not less than a few cm. The discussion will be limited to elastic scattering: that is, scattering of radiation of the same wavelength as the incident beam.

Consider a plane wave having wave vector $\boldsymbol{K}_0$ incident on a small piece of crystal and let us choose an arbitrary origin O within the crystal (Fig. 3.2). We assume that, because of the interaction between the wave and the

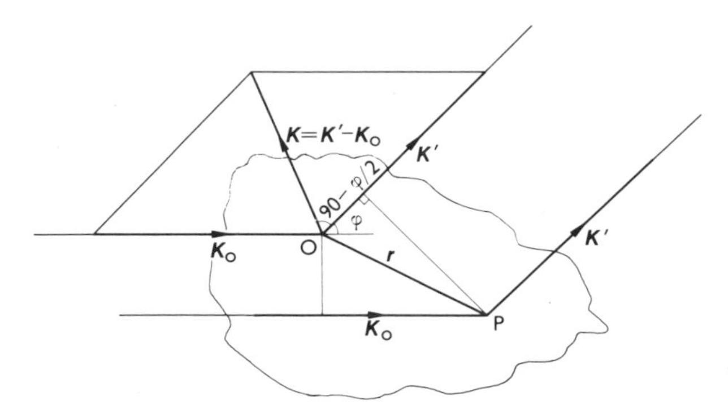

Figure 3.2 The relationship of the incident and scattered wave vectors $\boldsymbol{K}_0$ and $\boldsymbol{K}'$ to the scattering vector $\boldsymbol{K}$.

material of the crystal, each point within the crystal is the source of a spherical elastically scattered wave. The amplitudes of these spherical waves will depend upon the physical mechanism of the interaction but we may assume that they will be proportional to some function of the density of scattering matter located at their origins. Let us suppose that this function is $P(\boldsymbol{r})$, where $\boldsymbol{r}$ is the vector distance of the point in question from O.

We wish to find the resultant amplitude scattered with wave vector $\boldsymbol{K}'$ by the whole crystal. At distances far from O the phase difference between the radiation scattered in the direction of $\boldsymbol{K}'$ from O and from P (vector distance $\boldsymbol{r}$ away) is

$$\boldsymbol{r} \cdot \boldsymbol{K}_0 - \boldsymbol{r} \cdot \boldsymbol{K}' = \boldsymbol{r} \cdot (\boldsymbol{K}_0 - \boldsymbol{K}') = \boldsymbol{r} \cdot \boldsymbol{K} \qquad (3.1)$$

where $\boldsymbol{K}$ is the *scattering vector* and

$$\boldsymbol{K} = \boldsymbol{K}_0 - \boldsymbol{K}' = \frac{4\pi}{\lambda} \sin \frac{\phi}{2} \tag{3.2}$$

Thus, if the amplitude arriving at R from a volume element $d\tau$ at O is given by $A_0 P(0)\, d\tau$, that arriving from P is $A_0 P(\boldsymbol{r})\, e^{i\boldsymbol{K}.\boldsymbol{r}}\, d\tau_r$.

The total amplitude at R is obtained by integrating over the whole volume of crystal, when the *scattering amplitude* at R is given by

$$A_0 \int_{\text{crystal}} P(\boldsymbol{r})\, e^{i\boldsymbol{K}.\boldsymbol{r}}\, d\tau_r \tag{3.3}$$

The total scattered intensity at R is

$$I = I_0 \left| \int_{\text{crystal}} P(\boldsymbol{r})\, e^{i\boldsymbol{K}.\boldsymbol{r}}\, d\tau_r \right|^2 \tag{3.4}$$

3.3 The intensity diffracted by a lattice of point scatterers

Suppose now that the only scattering in the crystal arises from point scatterers, with scattering amplitude S, arranged on the corner points of a lattice defined by the basis vectors $\boldsymbol{a}$, $\boldsymbol{b}$ and $\boldsymbol{c}$. In this case, the integral of Equ. 3.4 can be replaced by a summation over all the lattice points. In this sum $\boldsymbol{r} = n_1\boldsymbol{a} + n_2\boldsymbol{b} + n_3\boldsymbol{c}$ where n_1, n_2 and n_3 are integers. If the crystal is a parallelepiped consisting of N_1, N_2 and N_3 cells in the $\boldsymbol{a}$, $\boldsymbol{b}$ and $\boldsymbol{c}$ directions respectively, then the intensity scattered with scattering vector $\boldsymbol{K}$ is given by

$$I = \left| \sum_{n_1=0}^{N_1} \sum_{n_2=0}^{N_2} \sum_{n_3=0}^{N_3} S\, e^{i\boldsymbol{K}.(n_1\boldsymbol{a} + n_2\boldsymbol{b} + n_3\boldsymbol{c})} \right|^2$$

$$= S^2 \left| \sum_{n_1=0}^{N_1} e^{i\boldsymbol{K}.n_1\boldsymbol{a}} \right|^2 \left| \sum_{n_2=0}^{N_2} e^{i\boldsymbol{K}.n_2\boldsymbol{b}} \right|^2 \left| \sum_{n_2=0}^{N_3} e^{i\boldsymbol{K}.(n_3\boldsymbol{c})} \right|^2.$$

Each of the terms in the product is the square of the sum of a geometric progression; the value of the first term is

$$\left| \frac{(1 - e^{iN_1\boldsymbol{K}.\boldsymbol{a}})}{1 - e^{i\boldsymbol{K}.\boldsymbol{a}}} \right|^2 = \frac{1 - \cos(N_1\boldsymbol{K}.\boldsymbol{a})}{1 - \cos(\boldsymbol{K}.\boldsymbol{a})} = \frac{\sin^2(\frac{1}{2}N_1\boldsymbol{K}.\boldsymbol{a})}{\sin^2(\frac{1}{2}\boldsymbol{K}.\boldsymbol{a})} \tag{3.5}$$

This term has a maximum value when $\frac{1}{2}\boldsymbol{K}.\boldsymbol{a}$ is an integral multiple of π. When this occurs, both numerator and denominator are zero and in the limit their ratio is $N_1{}^2$; Equ. 3.5 tends to zero when $\frac{1}{2}\boldsymbol{K}.\boldsymbol{a} = \left(m + \frac{1}{N_1}\right)\pi$,

m being an integer. It has subsequent subsidiary maxima whenever $\frac{1}{2}N_1\boldsymbol{K}\,.\,\boldsymbol{a}$ is an odd integral multiple of $\frac{\pi}{2}$. The sizes of the subsidiary maxima decrease rapidly and if $N_1 \ll n$ the height of the nth sub-maximum is $\frac{N_1^2}{\pi^2}\left(\frac{2}{2n+1}\right)^2$. Now the unit cell edges of most crystals are a few Ångström units (10^{-8} cm) in length, so that even in a very small crystal of thickness say 10^{-4} cm there will be at least 100 cells in each direction. There are thus at least 100 equally spaced maxima between each pair of main maxima. Since the heights of these are very small except very near the main maxima, having dropped to 1% by the seventh, it is a good approximation to equate the first term of Equ. 3.5 to zero unless $\boldsymbol{K}\,.\,\boldsymbol{a} = 2\pi h$ where h is an integer. Thus the total intensity, which is the product of three similar terms, will be essentially zero unless simultaneously

$$\boldsymbol{K}\,.\,\boldsymbol{a} = 2\pi h$$
$$\boldsymbol{K}\,.\,\boldsymbol{b} = 2\pi k$$
$$\boldsymbol{K}\,.\,\boldsymbol{c} = 2\pi l$$

where h, k and l are integers.

When these three conditions are satisfied the intensity is $N_1^2N_2^2N_3^2 = N^2$, where N is the total number of unit cells in the crystal.

3.4 The reciprocal lattice

In order to describe more easily the conditions for non-zero scattering it is convenient to introduce a *reciprocal lattice.* The values of $\boldsymbol{K}$ for which such scattering occurs can be defined by

$$\boldsymbol{K} = 2\pi(h\boldsymbol{a}^* + k\boldsymbol{b}^* + l\boldsymbol{c}^*) \tag{3.6}$$

where $\boldsymbol{a}^*$, $\boldsymbol{b}^*$ and $\boldsymbol{c}^*$ are basis vectors defining the reciprocal lattice such that

$$\boldsymbol{a}\,.\,\boldsymbol{a}^* = \boldsymbol{b}\,.\,\boldsymbol{b}^* = \boldsymbol{c}\,.\,\boldsymbol{c}^* = 1 \tag{3.7}$$

and

$$\boldsymbol{a}\,.\,\boldsymbol{b}^* = \boldsymbol{b}\,.\,\boldsymbol{c}^* = \boldsymbol{c}\,.\,\boldsymbol{a}^* = \boldsymbol{b}\,.\,\boldsymbol{a}^* = \boldsymbol{c}\,.\,\boldsymbol{b}^* = \boldsymbol{a}\,.\,\boldsymbol{c}^* = 0 \tag{3.8}$$

Thus the scattering vectors for which non-zero intensity is diffracted by a lattice of point scatterers are given by the radius vectors to points of the associated reciprocal lattice multiplied by 2π.

3.5 The Bragg equation

We now want to use the diffraction conditions given above to derive the familiar Bragg equation which expresses these conditions in terms of the distance between parallel crystallographic planes. Figure 3.3a shows a plane with Miller indices *hkl* intersecting the three crystallographic axes.

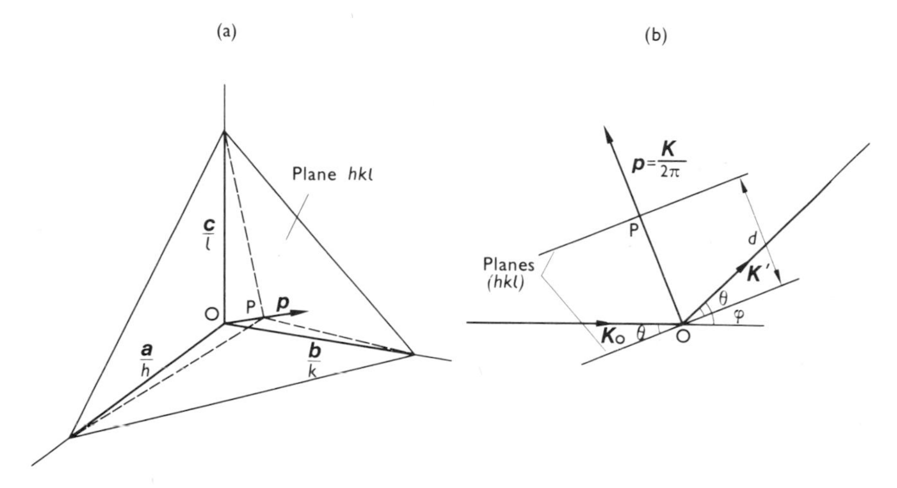

Figure 3.3 The geometrical conditions for Bragg reflection from a set of planes with Miller indices (*hkl*).

P is the base of the perpendicular from the origin O on to the plane and OP = d, the spacing between successive *hkl* planes. Let $\boldsymbol{p}$ be a vector parallel to OP passing through O, then, as in Equ. 1.2

$$\frac{\boldsymbol{a}\cdot\boldsymbol{p}}{h}=\frac{\boldsymbol{b}\cdot\boldsymbol{p}}{k}=\frac{\boldsymbol{c}\cdot\boldsymbol{p}}{l}=|\boldsymbol{p}|\,d \tag{3.9}$$

If we write $\boldsymbol{p}$ as a vector in the reciprocal lattice so that

$$\boldsymbol{p}=x\boldsymbol{a}^*+y\boldsymbol{b}^*+z\boldsymbol{c}^*$$

then Equ. 3.9 becomes (making use of Equ. 3.7 and 3.8)

$$\frac{x}{h}=\frac{y}{k}=\frac{z}{l}=|\boldsymbol{p}|d \tag{3.10}$$

A particularly simple solution of this equation occurs when $x = h$, $y = k$, $z = l$, then $|\boldsymbol{p}|d = 1$ and $\boldsymbol{p}$ is the reciprocal lattice vector $h\boldsymbol{a}^* + k\boldsymbol{b}^* + l\boldsymbol{c}^*$. The condition for strong diffraction with this reciprocal lattice vector is that

$$\boldsymbol{K} = 2\pi\boldsymbol{p} \tag{3.11}$$

The directions of incident and diffracted wave vectors for this value of $\boldsymbol{K}$ are shown in Fig. 3.3b, they are related as though by reflection in the planes hkl at glancing angle θ where $\theta = \phi/2$. Now from Equ. 3.1 and 3.11

$$2\pi\boldsymbol{p} = \frac{4\pi}{\lambda}\sin\theta$$

and combining this with Equ. 3.10 we obtain the *Bragg equation*

$$\lambda = 2d\sin\theta \tag{3.12}$$

One may note that other solutions of Equ. 3.10 than the one chosen describe higher orders of diffraction and would be given by the more general equation

$$n\lambda = 2d\sin\theta$$

It is however more convenient to describe the nth order diffraction from the planes with Miller indices (hkl) as first-order diffraction from planes (nh, nk, nl) and to retain the simple Bragg equation i.e. Equ. 3.12.

The reflection relationship between the incident and diffracted beams when the Bragg condition is satisfied has led to the widespread use of 'reflection' to describe the diffraction. The diffracted beam with scattering vector $2\pi(h\boldsymbol{a}^* + k\boldsymbol{b}^* + l\boldsymbol{c}^*)$ is commonly called 'the (hkl) reflection'. It is also worth pointing out that Equ. 3.11 shows that the reciprocal lattice vector $h\boldsymbol{a}^* + k\boldsymbol{b}^* + l\boldsymbol{c}^*$ is in the direction of the normal to the plane (hkl).

As was mentioned in Section 1.10, the spacing d of planes hkl may be expressed rather simply in terms of the reciprocal lattice vectors. The most general equation (for a triclinic cell) is

$$\left(\frac{1}{d}\right)^2 = d^{*2} = h^2\boldsymbol{a}^{*2} + k^2\boldsymbol{b}^{*2} + l^2\boldsymbol{c}^{*2} + 2kl\,\boldsymbol{b}^* \,.\, \boldsymbol{c}^* + \\ + 2lh\,\boldsymbol{c}^* \,.\, \boldsymbol{a}^* + 2hk\,\boldsymbol{a}^* \,.\, \boldsymbol{b}^* \tag{3.13}$$

3.6 Scattering from a real crystal

In a real crystal each unit cell contains, not a single point scatterer, but an assembly of atoms at positions within the cell defined by the crystal structure and having the symmetry of the space group to which the crystal belongs. The radius vector to any point in such a crystal may be written as

$$r = n_1 a + n_2 b + n_3 c + r_i + r_i'$$

where $n_1 a + n_2 b + n_3 c$ is the radius vector to the origin of the particular unit cell in which the point lies, r_i is the vector distance of the centre of the ith atom from the origin of the unit cell (we suppose that the point in question lies within the ith atom) and r_i' is the vector distance of the point from the centre of the ith atom. The integral of Equ. 3.3 can then be written

$$A = A_0 \sum_{n_1=0}^{N_1-1} \sum_{n_2=0}^{N_2-1} \sum_{n_3=0}^{N_3-1} \sum_{i=1}^{n} \int_{\text{atom}} P_i(r_i')\, e^{iK.(n_1a+n_2b+n_3c+r_i+r_i')}\, d\tau_{r_i'}$$

$$= A_0 \left[\sum_{n_1=0}^{N_1-1} \sum_{n_2=0}^{N_2-1} \sum_{n_3=0}^{N_3-1} e^{iK.(n_1a+n_2b+n_3c)} \right] \times$$

$$\sum_{i=1}^{n} \left(\int_{\text{atom}} P_i(r_i')\, e^{iK.r_i'}\, d\tau_{r_i'} \right) e^{iK.r_i} \qquad (3.14)$$

there being n atoms in the unit cell. The terms within the square brackets are identical with those arising in the case of point scatterers and show that the scattered amplitude is essentially zero unless K is 2π times a reciprocal lattice vector. The integral over the atom

$$\int_{\text{atom}} P_i(r_i')\, e^{iK.r_i'}\, d\tau_{r_i'} = f_i(K) \qquad (3.15)$$

is the *scattering factor* for the ith atom. Its magnitude and dependence upon K is given by the nature of the interaction between the atom and the incident wave which will be considered later. Inserting this into Equ. 3.14 and writing

$$r_i = x_i a + y_i b + z_i c \qquad (3.16)$$

and

$$K = 2\pi(ha^* + kb^* + lc^*)$$

we obtain

$$A = A_0 N \sum_{i=1}^{n} f_i(K)\, e^{2\pi i(hx_i + ky_i + lz_i)} \qquad (3.17)$$

the sum being taken over all the atoms in the unit cell. Thus the scattered intensity is given by

$$I = I_0 N^2 \left| \sum_{i=1}^{n} f_i(K)\, e^{2\pi i(hx_i + ky_i + lz_i)} \right|^2 \qquad (3.18)$$

It should be clear from this discussion that the intensity of the wave scattered elastically from a crystal can be evaluated in three stages. Firstly, the size and shape of the crystal lattice defines the directions in which scattering will occur, and these are most conveniently found using the reciprocal lattice. Secondly, the individual atoms in the unit cell are considered and the scattering due to each of these is represented by an appropriate atomic scattering factor or form factor. Lastly, the magnitudes of the form factors combined with the actual atomic positions in the unit cell enable the sum term in Equ. 3.17 to be evaluated: this sum is known as the *structure factor.* The values of the structure factors determine the relative intensities of the diffraction spectra permitted by the reciprocal lattice.

3.7 Systematically absent reflections

The value of the structure factor depends on the positions of atoms within the unit cell and these positions must show the symmetry of the space group to which the crystal belongs. If this space group has a non-primitive lattice, or contains one or more translational symmetry elements, then the phase relationships between the scattering from atoms related by these elements are such that for certain classes of reflection the structure factor is always zero. The way in which this arises is most easily seen by considering a few simple examples.

Body-centred (I) *lattice*

In a crystal having a body-centred (I) lattice, for every atom in the unit cell at a position (x_i, y_i, z_i) there is a related atom at $(\frac{1}{2}+x_i, \frac{1}{2}+y_i, \frac{1}{2}+z_i)$. From Equ. 3.17 the structure factor

$$\mathrm{F}(hkl) = \sum_{i=1}^{n} f_i(\boldsymbol{K})\, \mathrm{e}^{2\pi i(hx_i + ky_i + lz_i)}$$

and when the lattice is body-centred this can be written

$$F(hkl) = \sum_{i=1}^{n/2} f_i(\boldsymbol{K})\, (\mathrm{e}^{2\pi i(hx_i + ky_i + lz_i)} + \mathrm{e}^{2\pi i((\frac{1}{2}+x_i)h + (\frac{1}{2}+y_i)k + (\frac{1}{2}+z_i)l)})$$

$$= \sum_{i=1}^{n/2} f_i(\boldsymbol{K})[\mathrm{e}^{2\pi i(hx_i + ky_i + lz_i)}][1 + \mathrm{e}^{\pi i(h+k+l)}]$$

Table 3.1 Systematic absences introduced by non-primitive lattices and translational symmetry elements

	Description	Type of reflection affected	Condition for non-zero structure factor
Non-primitive lattices	A	hkl	
	B	hkl	
	C	hkl	
	F	hkl	h, k and l all even or all odd
	I	hkl	$h + k + l = 2n$
	R	$hkil$	$h + 2k + l = 3n$
Glide planes	$a \perp$ to $\mathbf{b}$	$(h0l)$	$h = 2n$
	$a \perp$ to $\mathbf{c}$	$(hk0)$	$h = 2n$
	$b \perp$ to $\mathbf{c}$	$(hk0)$	$k = 2n$
	$b \perp$ to $\mathbf{a}$	$(0kl)$	$k = 2n$
	$c \perp$ to $\mathbf{a}$	$(0kl)$	$l = 2n$
	$c \perp$ to $\mathbf{b}$	$(h0l)$	$l = 2n$
	$n \perp$ to $\mathbf{a}$	$(0kl)$	$k + l = 2n$
	$n \perp$ to $\mathbf{b}$	$(h0l)$	$h + l = 2n$
	$n \perp$ to $\mathbf{c}$	$(hk0)$	$h + k = 2n$
	$d \perp$ to $\mathbf{a}$	$(0kl)$	$k + l = 4n$
	$d \perp$ to $[1\bar{1}0]$	(hhl)	$2h + l = 4n$
Screw axes	$2_1 \parallel$ to $\mathbf{a}$	$(h00)$	$h = 2n$
	$2_1 \parallel$ to $\mathbf{b}$	$(0k0)$	$k = 2n$
	$2_1, 4_2, 6_3 \parallel$ to $\mathbf{c}$	$(00l)$	$l = 2n$
	$3_1, 3_2, 6_2, 6_4 \parallel$ to $\mathbf{c}$	$(00l)$	$l = 3n$
	$4_1, 4_3 \parallel$ to $\mathbf{c}$	$(00l)$	$l = 4n$
	$6_1, 6_5 \parallel$ to $\mathbf{c}$	$(00l)$	$l = 6n$

where the summation is taken over atoms not related by the body-centering. Now

$$1 + e^{\pi i(h+k+l)} = 1 + \cos \pi(h+k+l) + i \sin \pi(h+k+l)$$
$$= 1 + \cos \pi(h+k+l), \text{ since } h, k \text{ and } l \text{ are integers}$$
$$= 0 \text{ if } h+k+l \text{ is odd, or equal to 2 if } h+k+l \text{ is even.}$$

Thus the structure factor $F(hkl)$ will be zero unless $h + k + l$ is even.

b-glide plane perpendicular to **c**

If the unit cell contains a b-glide plane perpendicular to **c**, and the origin of the cell is chosen so that the glide plane passes through $z = 0$, then for every atom at x_i, y_i, z_i there will be a related atom $x_i, \frac{1}{2} + y_i, -z_i$. The structure factor then becomes

$$F(hkl) = \sum_{i=1}^{n/2} f_i(\boldsymbol{K})\,(e^{2\pi i(hx_i + ky_i + lz_i)} + e^{2\pi i(hx_i + k(\frac{1}{2} + y_i) - lz_i)})$$

where the summation is over those atoms not related by the glide plane

$$F(hkl) = \sum_{i=1}^{n/2} f_i(\boldsymbol{K})\,(e^{2\pi i(hx_i + ky_i + lz_i)} + e^{\pi ik}\, e^{2\pi i(hx_i + ky_i - lz_i)})$$

when $l = 0$ this becomes

$$F(hk0) = \sum_{i=1}^{n/2} f_i(\boldsymbol{K})\,(e^{2\pi i(hx_i + ky_i)})[1 + e^{\pi ik}]$$

The term in square brackets is zero when k is odd and hence if a crystal possesses a b-glide plane perpendicular to $\boldsymbol{c}$, reflections of the type $hk0$ are absent when k is odd.

Similar considerations to those given can be applied to deduce the systematic absences in other non-primitive lattices and for other types of translational symmetry element. The results obtained are summarised in Table 3.1.

3.8 Intensity diffracted by a uniformly rotating crystal

In the foregoing sections we have calculated the intensity scattered in a particular direction by a small parallelepiped of crystal. We have seen that this intensity is critically dependent on the relative orientation of the crystal and the incident wave vector. Now, since measurements of the structure factors must be made in order to determine the atomic positions, it is necessary to find some quantity which depends mainly on the structure factor and is not critically dependent on the crystal setting. Such a quantity is provided by the *integrated intensity.* Suppose a crystal is rotated with a uniform angular velocity ω about an axis perpendicular to $\boldsymbol{K}$ in a beam of incident intensity I_0. If E is the total power diffracted by the crystal as it sweeps through a Bragg reflection, then the integrated intensity of this Bragg reflection is defined as $E\omega/I_0$.

Now from Equ. 3.14 and 3.5, the intensity diffracted with scattering vector $\boldsymbol{K}$ is

$$I = I_0 |F(\boldsymbol{K})|^2 B^2(\boldsymbol{K}) \tag{3.19}$$

where $F(\boldsymbol{K})$ is the structure factor and $B^2(\boldsymbol{K})$ is the interference function

$$B^2(\boldsymbol{K}) = \frac{\sin^2 \frac{1}{2} N_1 \boldsymbol{K} \cdot \boldsymbol{a}}{\sin^2 \frac{1}{2} \boldsymbol{K} \cdot \boldsymbol{a}} \cdot \frac{\sin^2 \frac{1}{2} N_2 \boldsymbol{K} \cdot \boldsymbol{b}}{\sin^2 \frac{1}{2} \boldsymbol{K} \cdot \boldsymbol{b}} \cdot \frac{\sin^2 \frac{1}{2} N_3 \boldsymbol{K} \cdot \boldsymbol{c}}{\sin^2 \frac{1}{2} \boldsymbol{K} \cdot \boldsymbol{c}} \tag{3.20}$$

The power diffracted into a small range of solid angle $\mathrm{d}\Omega$ about $\boldsymbol{K}$ is

$$I \mathrm{d}\Omega = I_0 |F(\boldsymbol{K})|^2 B^2(\boldsymbol{K}) \, \mathrm{d}\Omega$$

Consider a set of rectangular Cartesian coordinates x, y and z in the $\boldsymbol{K}$-space of Fig. 3.4a with x parallel to $\boldsymbol{K}'$ and y in the plane of $\boldsymbol{K}_0$ and $\boldsymbol{K}'$. In terms of these coordinates it can be seen that

$$\mathrm{d}\Omega = \frac{\mathrm{d}y \, \mathrm{d}z}{|\boldsymbol{K}'|^2} = \frac{\lambda^2}{4\pi^2} \, \mathrm{d}y \mathrm{d}z$$

so that the total power diffracted when the incident wave vector is $\boldsymbol{K}_0$ is

$$P = I_0 \int_y \int_z |F(\boldsymbol{K})|^2 B^2(\boldsymbol{K}) \frac{\lambda^2}{4\pi^2} \, \mathrm{d}y \, \mathrm{d}z$$

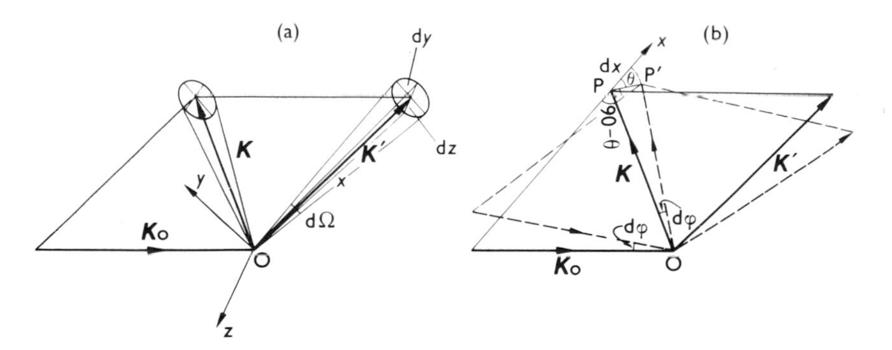

Figure 3.4 The orientation of the incident and scattered wave vectors and the scattering vector as a rotating crystal passes through the position for Bragg reflection.

As the crystal rotates there will be no significant scattering except at angles very close to $2\theta_0$ where θ_0 is the Bragg angle. The scattering vector $\boldsymbol{K}$ therefore moves through the same angle as the incident wave vector $\boldsymbol{K}_0$ as shown in Fig. 3.4b. In a small time increment $\mathrm{d}t$ the angle turned is

$d\phi(=\omega\, dt)$ and the total energy scattered during this time is $P\, dt = P(d\phi/\omega)$. From Fig. 3.4b it can be seen that

$$|\boldsymbol{K}\, d\phi| = |PP'| = \frac{dx}{\cos\theta_0}$$

$$\therefore d\phi = \frac{dx}{|\boldsymbol{K}|\cos\theta_0} = \frac{\lambda\, dx}{4\pi\sin\theta_0\cos\theta_0} = \frac{\lambda\, dx}{2\pi\sin 2\theta_0}$$

so that as the crystal rotates through the reflecting position the total energy scattered is

$$\int_x \frac{P}{\omega}\frac{\lambda\, dx}{2\pi\sin 2\theta_0} = \frac{I_0}{\omega}\int_x\int_y\int_z |F(\boldsymbol{K})|^2 B^2(\boldsymbol{K})\frac{\lambda^3}{8\pi^3\sin 2\theta_0}\, dxdydz \quad (3.21)$$

Now $dxdydz$ is a volume element in $\boldsymbol{K}$-space, and any $\boldsymbol{K}$-vector near to the direction of the hkl reciprocal lattice vector can be written

$$\boldsymbol{K} = 2\pi[(h+X)\boldsymbol{a}^* + (k+Y)\boldsymbol{b}^* + (l+Z)\boldsymbol{c}^*] \quad (3.22)$$

so that the volume element $dxdydz$ becomes $8\pi^3\boldsymbol{a}^*.(\boldsymbol{b}^*\times\boldsymbol{c}^*)\, dXdYdZ$. The interference function B only has an appreciable value when X, Y and Z in Equ. 3.22 are small, and since $F(\boldsymbol{K})$ is a slowly varying function of $\boldsymbol{K}$ we may replace it by $F(hkl)$ and take it outside the integration of Equ. 3.21. The integrated intensity can then be written

$$\frac{E\omega}{I_0} = |F(hkl)|^2\frac{\lambda^3}{V\sin 2\theta_0}\int_X\int_Y\int_Z B\, dXdYdZ \quad (3.23)$$

where V is the volume of the unit cell, since from Equ. 3.7 and 3.8

$$\boldsymbol{a}^*.(\boldsymbol{b}^*\times\boldsymbol{c}^*) = \frac{1}{\boldsymbol{a}.(\boldsymbol{b}\times\boldsymbol{c})} = \frac{1}{V}$$

The integration of Equ. 3.23 is separable into three similar terms when the value of B is substituted from Equ. 3.20. The first of these is

$$\int_X \frac{\sin^2\pi N_1\boldsymbol{K}.\boldsymbol{a}}{\sin^2\pi\boldsymbol{K}.\boldsymbol{a}}\, dX = \int_X \frac{\sin^2\pi N_1(h+X)}{\sin^2\pi(h+X)}\, dX = \int_X \frac{\sin^2\pi N_1 X}{\sin^2\pi X}\, dX$$

Since the integrand is only different from zero when X is small, we may take the limits of integration as $+\infty$ and $-\infty$. When X is small, $\sin^2\pi X = (\pi X)^2$ and the integral can be approximated as

$$\int_{-\infty}^{\infty}\frac{\sin^2\pi N_1 X}{(\pi X)^2}\, dX = N_1\int_{-\infty}^{\infty}\frac{\sin^2 U}{U^2}\, dU = N_1 \quad (3.24)$$

and thus the integrated intensity is given by

$$\frac{E\omega}{I_0} = |F(hkl)|^2 \frac{\lambda^3}{V \sin 2\theta_0} = |F(hkl)|^2 \frac{\lambda^3}{\sin 2\theta_0} n^2 v = Qv \qquad (3.25)$$

where n is the number of unit cells per unit volume and v is the total volume of the crystal. The *integrated intensity per unit volume of crystal* is commonly referred to as Q.

3.9 The Fourier transform

It is convenient to introduce here the concept of the *Fourier transform.* (A brief outline of the properties of this transform are given in the Appendix for readers not already familiar with it.) In Section 3.2 we showed that the amplitude scattered by a crystal with scattering vector **K** is

$$A(\boldsymbol{K}) = \int_{-\infty}^{\infty} P(\boldsymbol{r})\, e^{i\boldsymbol{K}.\boldsymbol{r}}\, d\tau_r$$

i.e., a volume integral in real space.

The equation simply represents a Fourier transformation relationship between the functions $A(\boldsymbol{K})$ and $P(\boldsymbol{r})$ so that, applying Fourier's integral theorem, we obtain

$$P(\boldsymbol{r}) = \int_{-\infty}^{\infty} A(\boldsymbol{K})\, e^{-i\boldsymbol{K}.\boldsymbol{r}}\, d\tau_K$$

i.e., a volume integral in $\boldsymbol{K}$-space.

Now we found previously that $A(\boldsymbol{K}) = F(\boldsymbol{K})B(\boldsymbol{K})$ where $F(\boldsymbol{K})$ is the structure factor and $B(\boldsymbol{K})$ is the amplitude of the interference function. Then, since $B(\boldsymbol{K})$ is small unless $\boldsymbol{K} = 2\pi$ times a reciprocal lattice vector and $F(\boldsymbol{K})$ is essentially constant over the ranges of $\boldsymbol{K}$ for which $B(\boldsymbol{K})$ is non-zero, we can write

$$\boldsymbol{K} = ((h + X)\boldsymbol{a}^* + (k + Y)\boldsymbol{b}^* + (l + Z)\boldsymbol{c}^*)2\pi$$

$$\boldsymbol{r} = x\boldsymbol{a} + y\boldsymbol{b} + z\boldsymbol{c}$$

$$d\tau_K = dXdYdZ\, (\boldsymbol{a}^* . (\boldsymbol{b}^* \times \boldsymbol{c}^*)) = \frac{dXdYdZ}{V}$$

so that $P(\boldsymbol{r}) = \sum_{h=-\infty}^{\infty} \sum_{k=-\infty}^{\infty} \sum_{l=-\infty}^{\infty} \frac{F(hkl)}{V} \iiint B(XYZ)\, dXdYdZ$

where the integration is to be carried out around one reciprocal lattice point.

Now $B(XYZ)$ is the product of three terms of the form

$$B(X) = \frac{1 - \exp(2\pi i N_1 X)}{1 - \exp(2\pi i X)}$$

and when N_1 is large the value of the integral approaches unity. Thus

$$P(xyz) = \frac{1}{V} \sum_{-\infty \atop h}^{\infty} \sum_{-\infty \atop k}^{\infty} \sum_{-\infty \atop l}^{\infty} F(hkl) \tag{3.26}$$

From this equation we conclude that if the structure factors of all reflections could be measured it would be possible to deduce $P(xyz)$ uniquely. Unfortunately this can never be achieved, because reflections corresponding to all possible combinations hkl can never be observed using radiation of finite wavelength. For, since

$$\lambda = 2d \sin\theta = \frac{4\pi}{K} \sin\theta$$

and the maximum value of $\sin\theta = 1$, $K_{max.} = 4\pi/\lambda$. Thus the summation for $P(\boldsymbol{r})$ is perforce terminated at reflections with $\boldsymbol{K}$ values of $4\pi/\lambda$. The effect of this termination of the series is to limit the resolution with which $P(\boldsymbol{r})$ can be determined. Since the Fourier coefficients with periodic lengths less than $2\pi/K_{max.} = \lambda/2$ are omitted, variations of $P(\boldsymbol{r})$ over distances shorter than this cannot be deduced from the summation. For maximum resolution a short wavelength is used and reflections observed out to as high an angle as possible.

3.10 The phase problem

The second major difficulty in deriving the distribution $P(\boldsymbol{r})$ from the structure factors is that these cannot as yet be measured directly in an experiment. The quantity which is usually measured is the integrated intensity (§3.8) and this is proportional to the square of the structure factor. In general, the structure factor is a complex quantity so that all that can be determined from the measurement of integrated intensity is its amplitude, its phase being unknown. The problem of determining the phases of the structure factors is usually the major difficulty in a structure determination. There are very many methods which can be used in attempts to solve it; however it would not be appropriate to give an account of such methods here. Suffice it to say that it is usually possible to deduce an approximate structure for a crystal using such methods and taking due account of the symmetry, the sizes of the atoms and any details of the chemical configuration which may be known. If the structure amplitudes calculated for this trial structure are in fair agreement with those deduced from the integrated intensities, then the phases calculated from the trial structure can be used with the measured structure amplitudes to obtain the distribution $P(xyz)$ and hence a better approximation to the structure.

It is worth noticing that if the structure has a centre of symmetry the phase problem is greatly simplified. Suppose a certain structure has a centre of symmetry: if we choose the origin of the unit cell to be coincident with this centre, then every atom at $x_i y_i z_i$ is related by this centre to a similar atom at $\bar{x}_i \bar{y}_i \bar{z}_i$. The structure factor

$$F(hkl) = \sum_{i=1}^{n} f_i \, e^{2\pi i(hx_i + ky_i + lz_i)}$$

can now be written

$$F(hkl) = \sum_{i=1}^{n/2} f_i \, [e^{2\pi i(hx_i + ky_i + lz_i)} + e^{-2\pi i(hx_i + ky_i + lz_i)}]$$

$$= 2 \sum_{i=1}^{n/2} f_i \cos 2\pi(hx_i + ky_i + lz_i)$$

where the sum is to be taken over all atoms in the unit cell not related by the centre of symmetry. This shows that, when a centre of symmetry is present, then as long as the scattering factors f_i of all the atoms are real, the structure factor will be real and the phase problem is reduced to one of sign determination.

3.11 Friedel's law

It was mentioned in an earlier chapter (§ 1.6) that in most cases there is no difference in the intensities scattered by the planes hkl and $\overline{hkl}$, even though the crystal may not possess a centre of symmetry. The truth of this statement may be seen by examining the form of the structure factors for the two reflections

$$F(hkl) = \sum_i f_i \exp 2\pi i(hx + ky + lz)$$

$$= \sum_i f_i \cos 2\pi(hx + ky + lz) + i \sum_i f_i \sin 2\pi(hx + ky + lz)$$

similarly $F(\overline{hkl}) = \sum_i f_i \cos 2\pi(-hx - ky - lz) + i \sum_i f_i \sin 2\pi(-hx - ky - lz)$

$$= \sum_i f_i \cos 2\pi(hx + ky + lz) - i \sum_i f_i \sin 2\pi(hx + ky + lz)$$

The measured intensity is proportional to $|F(hkl)|^2$ which can be seen to be the same for the two cases so long as the atomic scattering factor f_i are real. The equivalence of the intensities of reflections $\overline{hkl}$ and hkl, unless anomalous scattering is present, means that it it usually impossible to determine directly from the symmetry of the scattering whether or not a crystal possesses a centre of symmetry. This indeterminacy is known as *Friedel's law.*

3.12 Scattering from disordered systems

Many crystals which are encountered in practice do not have perfectly ordered structures; a simple and important example is a random solid solution. In such a disordered crystal, the positions which the atoms occupy may be perfectly regular, but these positions are occupied randomly by one or the other component of the solid solution. Since our derivation of the scattering formulae has assumed that each set of equivalent positions is occupied by a single atomic species, it is of interest to see to what extent the same formulae will be valid for disordered structures. To do this we define a mean atomic scattering factor for each set of equivalent positions and let this be $\bar{f}_j$ for the jth set in the cell. Then we can write the scattering factor for the ith atom of the jth set in the nth cell as

$$f_{ijn} = \bar{f}_j + \Delta f_{ijn}$$

where the Δf_{ijn} are the differences of the scattering factors of atoms at each of the positions from the mean. Thus

$$\sum_j \sum_n \Delta f_{ijn} = 0 \quad \text{and also} \quad \sum_n \Delta f_{ijn} = 0$$

if the disorder is completely random.

The scattered amplitude $A(\boldsymbol{K})$ then becomes

$$A(\boldsymbol{K}) = \sum_n \sum_j \bar{f}_j \, e^{i\boldsymbol{K}\,.\,(r_{nij})} + \sum_n \sum_j \sum_i \Delta f_{ijn} \, e^{i\boldsymbol{K}\,.\,(r_{nij})}$$

When $\boldsymbol{K} = 2\pi(h\boldsymbol{a}^* + k\boldsymbol{b}^* + l\boldsymbol{c}^*)$, $\quad e^{i\boldsymbol{K}\,.\,\mathbf{r}_n} = 1$

and $A(\boldsymbol{K}) = NF(hkl) + \sum_j \sum_i e^{i\boldsymbol{K}\,.\,\mathbf{r}_{ij}} \sum_n \Delta f_{ijn}$

$$= NF(hkl)$$

since the second term is zero. In this case,

$$F(hkl) = \sum_i \sum_j \bar{f}_i \, e^{i\boldsymbol{K}\,.\,r_{ij}}.$$

When $\boldsymbol{K} \neq 2\pi(h\boldsymbol{a}^* + k\boldsymbol{b}^* + l\boldsymbol{c}^*)$ the first term is zero and

$$A(\boldsymbol{K}) = \sum_n \sum_i \Delta f_{in} \, e^{i\boldsymbol{K}.\mathbf{r}_{ni}}$$

Since the disorder is random, the values of Δf_{ijn} are completely uncorrelated with the particular values of j and n. The contributions $\Delta f_{ijn} \, e^{i\boldsymbol{K}.r_{nij}}$ to the sum over n (the cells of the crystal) and i (the atoms in an equivalent set) can be represented on an amplitude–phase diagram by a series of vectors of length Δf_{in} in random orientations to one another. This is an equivalent situation to the random walk problem where Δf_{in} is the step length; the resultant amplitude, being the mean distance travelled in n steps is $(\overline{(\Delta f_{in})^2 N})^{1/2}$. If all the equivalent positions are subject to random disorder and $\overline{\Delta f_i^2}$ is the mean-square deviation of the scattering factors from the mean of the ith atom of the cell, then the scattering when $\boldsymbol{K}$ is not 2π times a reciprocal lattice vector is $|A(\boldsymbol{K})|^2 = N \sum_i \overline{\Delta f_i^2}$.

We may conclude that when random disorder is present in a crystal, the scattering can be considered in two parts; the first part appears in the same positions as would be observed for an ordered structure, the appropriate structure amplitudes being calculated using the mean atomic scattering factor appropriate to each site in the unit cell. The second part, which has no equivalent in an ordered structure, is zero at the Bragg reflections but elsewhere has the intensity $N \sum_i \overline{\Delta f_i^2}$. This part, which is known as *incoherent scattering*, is independent of crystal orientation and only depends upon $\boldsymbol{K}$ through variations in the scattering factors with angle which affect $\overline{\Delta f_i^2}$; it therefore appears as a more-or-less uniform background in the diffraction pattern.

Let us take as an example a simple binary alloy having a body-centred cubic structure and containing a fraction x of atoms with scattering factor A and a fraction $(1 - x)$ with scattering factor B. The mean scattering factor for such an alloy is $Ax + B(1 - x)$ and the differences of the scattering factors from the mean are

$$A - B - Ax + Bx = (A - B)(1 - x)$$

for sites occupied by A atoms, and

$$B - B - Ax + Bx = (B - A)x$$

for sites occupied by B atoms. Hence the mean-square deviation of the scattering factors is

$$(A - B)^2 [(1 - x)^2 x + x^2(1 - x)] = (A - B)^2 (x - x^2)$$

so that the incoherent intensity would be $N(A - B)^2 (x - x^2)$.

3.13 The effect of temperature on the scattering

A further important departure from complete regularity in a crystal arises from the oscillations, due to thermal energy, of the atoms about their mean positions. A rigorous treatment of the scattering from a lattice containing oscillating atoms would require a detailed knowledge of the phonon spectrum of the crystal and is far beyond the scope of this book. For further details, the reader should consult *The Dynamics of Atoms in Crystals* by W. Cochran.

However it is possible, by considering the thermal oscillations of the atoms to be a form of random disorder, to use the results of the previous section to show how the coherent scattering from such a lattice differs from that of the lattice containing stationary atoms considered up to now.

In this analysis we will assume that the thermal motions of the individual atoms are independent of one another. This assumption is certainly not justified in real crystals, but gives results in good agreement with experiment if we consider the elastic coherent scattering only. We will also assume that the oscillations are isotropic and harmonic, an assumption that is good to a first approximation for relatively symmetrical crystals. Under these conditions, the probability that an atom will be displaced by a distance r_t from its mean position can be written

$$P(r_t) = N \exp - (r_t{}^2/2\overline{U^2})$$

where N is a normalising constant and $\overline{U^2}$ is the root-mean-square displacement of an atom from its equilibrium position. The atomic scattering factor for an atom is given by Equ. 3.15 as

$$f(\boldsymbol{K}) = \int_{\text{atom}} P(\boldsymbol{r})\, e^{i\boldsymbol{K}\cdot\boldsymbol{r}}\, d\tau_r$$

The scattering factor referred to the same origin for a similar atom displaced a vector distance $\boldsymbol{r}_t$ from that origin would be

$$f'(\boldsymbol{K}) = \int_{\text{atom}} P(\boldsymbol{r})\, e^{i\boldsymbol{K}\cdot(\boldsymbol{r}+\boldsymbol{r}_t)}\, d\tau_r$$

where $\boldsymbol{r}$ is measured from the new atomic centre. Thus the mean atomic scattering factor for an oscillating atom may be written

$$\bar{f}(\boldsymbol{K}) = \int_{\substack{\text{atom}\\ r}} \int_{\substack{-\infty\\ r_t}}^{\infty} P(\boldsymbol{r})P(\boldsymbol{r}_t)\, e^{i\boldsymbol{K}\cdot(\boldsymbol{r}+\boldsymbol{r}_t)}\, d\tau_r\, d\tau_{r_t} \Bigg/ \int_{-\infty}^{\infty} P(\boldsymbol{r}_t)\, d\tau_{r_t}$$

The two integrals in the numerator are separable so that

$$\bar{f}(\boldsymbol{K}) = \frac{\int_{\text{atom}} P(\boldsymbol{r})\, e^{i\boldsymbol{K}.\boldsymbol{r}}\, d\tau_r \int_{-\infty}^{\infty} N\, e^{-r_t^2/2\overline{U^2}}\, e^{i\boldsymbol{K}.\boldsymbol{r}_t}\, d\tau_{rt}}{\int_{-\infty}^{\infty} N\, e^{-r_t^2/2\overline{U^2}}\, d\tau_{rt}}$$

the integral over $\boldsymbol{r}$ is just the atomic scattering factor $f(\boldsymbol{K})$ and the integrals over $\boldsymbol{r}_t$ can be evaluated by writing $\boldsymbol{r}_t$ and $\boldsymbol{K}$ in their components r_x, r_y, r_z, K_x, K_y, K_z in a system of rectangular Cartesian coordinates. The volume element $d\tau_{rt}$ then becomes $dr_x\, dr_y\, dr_z$ and the integrals are the products of three similar terms of which the first is

$$\int_{-\infty}^{\infty} e^{-r_x^2/2\overline{U^2}}\, e^{iK_x r_x}\, dr_x \Big/ \int_{-\infty}^{\infty} e^{-r_x^2/2\overline{U^2}}\, dr_x$$

the integrals of both numerator and denominator can be evaluated for these particular limits, and the result is

$$\sqrt{2\pi\overline{U^2}}\; e^{-K_x^2\overline{U^2}/2} \Big/ \sqrt{2\pi\overline{U^2}}$$

Hence the mean scattering factor $\bar{f}(\boldsymbol{K})$ becomes

$$\bar{f}(\boldsymbol{K}) = f(\boldsymbol{K}) \exp - (\tfrac{1}{2}\overline{U^2}\,(K_x^2 + K_y^2 + K_z^2))$$
$$= f(\boldsymbol{K}) \exp - \left(\frac{K^2\overline{U^2}}{2}\right)$$

The result of the atomic vibration is therefore to reduce the atomic scattering factor for each atom by the factor $\exp - (K^2\overline{U^2}/2)$, where $\overline{U^2}$ is the mean-square displacement of the atom concerned from its equilibrium position. Since $K = 4\pi \sin\theta/\lambda$, this exponential factor can be written e^{-BS^2} where $B = 8\pi^2\overline{U^2}$ and $S^2 = (\sin^2\theta)/\lambda^2$. The factor B is known as the *temperature factor* of the atom since B is directly dependent on $\overline{U^2}$; it increases as the temperature is increased and the thermal vibration becomes greater in amplitude. Typical values of B at room temperature are in the range 0·2–0·8 Å^2. The exponential factor depends upon $\sin^2\theta/\lambda^2$ and is therefore nearly unit at low angles, hence the low-angle scattering is relatively insensitive to thermal vibration. At higher angles the exponential factor falls quite significantly from unity: at $\sin\theta/\lambda = 1$ for $B = 0{\cdot}5$ Å^2 the atomic scattering factor is reduced by a factor of about 0·6 from its value for a stationary atom.

4

The Scattering of X-rays, Electrons and Neutrons

4.1 Introduction

In order to recognise the useful fields of application of the different types of radiation which are used to study the structure of crystals, it is necessary to take some account of the physical mechanisms involved in the scattering processes.

4.2 X-Ray scattering

X-Ray scattering arises from the electromagnetic interaction between the electric field vector of the electromagnetic wave (X-ray) and the electrons in the crystal. The interaction induces a time-dependent fluctuation of the wavefunction of the electrons in the crystal and the fluctuating current density corresponding to these changes is the source of the scattered waves. The component which has the same frequency as the incident wave gives rise to coherent Bragg scattering and this is the only part that will be considered here. The intensity of such scattering can be calculated from first-order perturbation theory using a semi-classical treatment in which the electrons are treated quantum mechanically but the radiation field is not quantised. For an electron of mass m in a field of scalar potential ϕ and vector potential $\boldsymbol{A}$ the Hamiltonian is written

$$\mathcal{H} = \frac{1}{2m}\left(\boldsymbol{P} - \frac{e\boldsymbol{A}}{c}\right)^2 + e\phi$$

and if the vector potential $\boldsymbol{A}$ is small so that it can be treated as a small perturbation

$$\mathcal{H} = \frac{1}{2m}P^2 - \frac{e}{mc}\boldsymbol{A}\,.\,\boldsymbol{P} + e\phi$$

Writing the momentum operator $\boldsymbol{P}$ as $(\hbar/i)\nabla$, the time-dependent Schrödinger equation is

$$-\frac{\hbar^2}{2m}\nabla^2\psi - \frac{e\hbar}{imc}\boldsymbol{A}\,.\,\nabla\psi + e\phi\psi = i\hbar\frac{\partial\psi}{\partial t} \tag{4.1}$$

If we multiply this equation by ψ^* and the conjugate equation by ψ and subtract the two we obtain

$$-\frac{\hbar^2}{2m}(\psi^*\nabla^2\psi - \psi\nabla^2\psi^*) - \frac{\hbar e\boldsymbol{A}}{imc}\,.\,(\psi\nabla\psi^* - \psi^*\nabla\psi) = i\hbar\frac{\partial}{\partial t}(\psi\psi^*)$$

or $$\frac{i\hbar e}{2m}\nabla\,.\,(\psi\nabla\psi^* - \psi^*\nabla\psi) + \frac{e^2}{mc}(\psi\psi^*\,\nabla\,.\,\boldsymbol{A} - \nabla\,.\,\psi\psi^*\boldsymbol{A}) = \frac{\partial}{\partial t}(e\psi\psi^*)$$

But for a plane wave $\nabla\,.\,\boldsymbol{A} = 0$, so

$$\nabla\,.\left(\frac{i\hbar e}{2m}(\psi\nabla\psi^* - \psi^*\nabla\psi) - \frac{e^2}{mc}\psi\psi^*\boldsymbol{A}\right) = \frac{\partial}{\partial t}(e\psi\psi^*) \tag{4.2}$$

Comparing this with the classical equation of continuity

$$\nabla\,.\,[\boldsymbol{J}] = -\frac{1}{c}\frac{\partial\rho}{\partial t} \tag{4.3}$$

and remembering that $e\psi\psi^*$ represents a charge density, it can be seen that the quantum mechanical expression for the current density is

$$\boldsymbol{J} = -\frac{1}{c}\left(\frac{i\hbar e}{2m}(\psi\nabla\psi^* - \psi^*\nabla\psi) - \frac{e^2}{mc}\boldsymbol{A}\psi\psi^*\right) \tag{4.4}$$

The vector potential at a point outside the crystal, vector distance $\boldsymbol{R}$ from the origin, caused by a current density $\boldsymbol{J}$ is given by classical radiation theory as

$$\boldsymbol{A}_{\mathrm{S}} = \int_{\text{crystal}} \frac{[\boldsymbol{J}]\,\mathrm{d}\tau_r}{|\boldsymbol{r}+\boldsymbol{R}|}$$

where the square brackets round $\boldsymbol{J}$ denote the retarded value of the current density. The vector potential is related to the electric field by the equation

$$\boldsymbol{E} = -\frac{1}{c}\frac{\partial \boldsymbol{A}}{\partial t} - \nabla\phi$$

and hence $$\boldsymbol{E}_{\mathrm{S}} = -\frac{1}{c}\int_{\text{crystal}}\left[\frac{\partial[\boldsymbol{J}]}{\partial t}\frac{1}{|\boldsymbol{r}+\boldsymbol{R}|}\right]\mathrm{d}\tau_r - \nabla\phi \tag{4.5}$$

Thus the amplitude of the scattered radiation $\boldsymbol{E}_S$ depends upon the rate of change of the induced current density. Under the action of the perturbing field the wavefunction of Equ. 4.4 will change. However, if the perturbation is small and the frequency of the radiation is far from any of the electron transition frequencies of the atoms in the crystal, then the change in wavefunction will be insignificant compared to the oscillation of the vector potential $\boldsymbol{A}$. Hence to a good approximation we can write

$$\frac{\mathrm{d}\boldsymbol{J}}{\mathrm{d}t} = \frac{e^2}{mc^2}\frac{\partial \boldsymbol{A}}{\partial t}(\psi\psi^*)$$

and $$\boldsymbol{E}_S = \frac{e^2}{mc^2}\int_{\text{crystal}}\left[([\boldsymbol{E}]\psi\psi^* + \nabla\phi)\frac{1}{|\boldsymbol{r}+\boldsymbol{R}|}\right]\mathrm{d}\tau_r - \nabla\phi$$

where $[\boldsymbol{E}]$ is the retarded value of the incident electric field.

Now if we assume that $\boldsymbol{R} \gg \boldsymbol{r}$ we can write,

$$[\boldsymbol{E}] = \boldsymbol{E}_0\,\mathrm{e}^{i(\omega t - \boldsymbol{r}\,.\,\boldsymbol{K}_0 - \boldsymbol{r}\,.\,\boldsymbol{K}' - \boldsymbol{R}\,.\,\boldsymbol{K}')}$$

and

$$\boldsymbol{E}_S = \frac{e^2}{mc^2}\frac{1}{R}\boldsymbol{E}_0\,\mathrm{e}^{i(\omega t - \boldsymbol{K}'\,.\,\boldsymbol{R})}\int_{\text{crystal}}[\psi\psi^*\,\mathrm{e}^{i\boldsymbol{K}\,.\,\boldsymbol{r}}]\,\mathrm{d}\tau_r + \frac{e^2}{mc^2}\int_{\text{crystal}}\left[\nabla\phi\frac{1}{|\boldsymbol{R}+\boldsymbol{r}|}\right]\mathrm{d}\tau_r - \nabla\phi$$

Only the first of these three terms has the frequency of the incident radiation, hence only this term contributes to the elastically scattered wave. The field $\boldsymbol{E}_S$ is parallel to the incident electric field amplitude E_0, but only its components perpendicular to the direction of $\boldsymbol{K}'$ will contribute to scattering with this wave vector. Since the angle between $\boldsymbol{K}_0$ and $\boldsymbol{K}'$ is twice the Bragg angle θ, the appropriate components will have magnitude $\boldsymbol{E}_S$ if $\boldsymbol{E}_0$ is perpendicular to both $\boldsymbol{K}$ and $\boldsymbol{K}'$, and magnitude $\boldsymbol{E}_S \cos 2\theta$ if $\boldsymbol{E}_0$ lies in the plane of $\boldsymbol{K}$ and $\boldsymbol{K}'$. For unpolarised incident radiation the appropriate factor is $\frac{1}{2}(1 + \cos 2\theta)\boldsymbol{E}_S$. Thus the amplitude of the elastically scattered radiation can be written

$$\boldsymbol{E}_{S0} = \tfrac{1}{2}(1 + \cos 2\theta)\frac{e^2}{mc^2}\int_{\text{crystal}}\psi\psi^*\,\mathrm{e}^{i\boldsymbol{K}\,.\,\boldsymbol{r}}\,\mathrm{d}\tau_r \qquad (4.6)$$

Comparing this with Equ. 3.3 we see that for scattering of unpolarised X-radiation

$$P(\boldsymbol{r}) = \tfrac{1}{2}(1 + \cos 2\theta)\frac{e^2}{mc^2}\psi\psi^* \qquad (4.7)$$

and the atomic scattering factor for X-ray scattering is

$$f_X(\boldsymbol{K}) = \tfrac{1}{2}(1 + \cos 2\theta)\frac{e^2}{mc^2}\int_{\text{atom}} \psi\psi^* \, e^{i\boldsymbol{K}.\boldsymbol{r}} \, d\tau_r \tag{4.8}$$

In order to calculate this atomic scattering factor the electron distribution $\psi\psi^*$ within each atom must be known. The electron distributions, and hence the scattering factors, have been calculated for most atoms using modified Hartree–Fock self-consistent field methods. These scattering factors are tabulated in the *International Tables for Crystallography*, Volume III, (1962). The values given are sufficiently precise to be used in all but the most accurate work. Alternatively, in a crystal of known simple structure the scattering measurements themselves may be used to deduce scattering factors and hence to determine the electron distribution in the atoms of the crystal.

4.3 Electron scattering

The wavelength of an electron with momentum p is given by the De Broglie relationship $\lambda p = h$. In electron-diffraction experiments the electrons are commonly accelerated through potentials of about 100 kV and hence their energy $E = p^2/2m = eV = 10^5$ electron volts $= 1{\cdot}6 \times 10^{-7}$ erg. Their wavelength λ is given by

$$\lambda = \frac{h}{\sqrt{2mE}} = \frac{2\pi \times 10^{-27}}{1{\cdot}7 \times 10^{-18}}$$

$$\approx 4 \times 10^{-9} \text{ cm} = 0{\cdot}4 \text{ Å}$$

This is of the same order of magnitude as the distance between atoms of the crystal so that the conditions for diffraction are satisfied. An exact calculation of the electron wavelength involves use of the relativistic relationship between the energy and momentum of the electron; the error involved is about 5% at 100 kV and rises steadily for higher voltages. The interaction involved in electron scattering is the straightforward coulomb force between the incident electron and the electrons and nucleus of the atom. Now the conditions that we assumed in Section 3.2 are just those required for the validity of the Born approximation for particle scattering (see Mott and Massey, 1949). In this case the wave function for elastically scattered electrons is given by

$$\psi_s = \frac{m_0 e}{2\pi\hbar^2}\frac{e^{i\boldsymbol{K}_0.\boldsymbol{R}}}{|\boldsymbol{R}|}\int_{\text{crystal}} \phi(\boldsymbol{r})\, e^{i\boldsymbol{K}.\boldsymbol{r}} \, d\tau_r \tag{4.9}$$

where m_0 is the rest mass of the electron, $\boldsymbol{R}$ is the vector distance from the origin to the point at which the wave function is ψ_s, and $\phi(\boldsymbol{r})$ is the electrostatic potential in the crystal. Comparing Equ. 4.9 with Equ. 3.3 shows that in the case of electron scattering the function $P(\boldsymbol{r})$ may be written

$$P(\boldsymbol{r}) = \frac{m_0 e}{2\pi\hbar^2}\,\phi(\boldsymbol{r}) \tag{4.10}$$

The atomic scattering factor for electrons is

$$f_e(\boldsymbol{K}) = \frac{m_0 e}{2\pi\hbar^2} \int_{\text{atom}} \phi(\boldsymbol{r})\, e^{i\boldsymbol{K}\cdot\boldsymbol{r}}\, d\tau_r \tag{4.11}$$

Within an atom the potential $\phi(\boldsymbol{r})$ may be written

$$\phi(\boldsymbol{r}) = \int_{\text{atom}} \frac{\rho(\boldsymbol{r}')}{|\boldsymbol{r}-\boldsymbol{r}'|}\, d\tau_{r'}$$

where $\rho(\boldsymbol{r}')$ is the charge density at the point with radius vector $\boldsymbol{r}'$, so that

$$f_e(\boldsymbol{K}) = \frac{m_0 e}{2\pi\hbar^2} \int \rho(\boldsymbol{r}')\, e^{-i\boldsymbol{K}\cdot\boldsymbol{r}'}\, d\tau_{r'} \int \frac{1}{|\boldsymbol{r}-\boldsymbol{r}'|}\, e^{i\boldsymbol{K}\cdot(\boldsymbol{r}-\boldsymbol{r}')}\, d\tau_r \tag{4.12}$$

The first integral is $e(Z - \int \psi\psi^*\, e^{i\boldsymbol{K}\cdot\boldsymbol{r}}\, d\tau_r)$ where Z is the nuclear charge; it should be noticed that the integral required is just that involved in X-ray scattering (Equ. 4.6). The second integral can be evaluated between the limits zero and infinity and the result is $4\pi/K^2$. The electron scattering factor then becomes

$$f_e(\boldsymbol{K}) = \frac{2me^2}{\hbar^2 K^2}\left(Z - \int_{\text{atom}} \psi\psi^*\, e^{i\boldsymbol{K}\cdot\boldsymbol{r}}\, d\tau_r\right) \tag{4.13}$$

The second term within the bracket is closely related to the X-ray scattering factor (Equ. 4.8) and the electron scattering factor is thus easily evaluated if the X-ray scattering factor is known. It should be noted that the scattering factor for $\boldsymbol{K} = 0$ cannot be obtained from Equ. 4.13 since both numerator and denominator tend to zero. Integration of Equ. 4.12 for the special case $\boldsymbol{K} = 0$ shows that the scattering factor has a maximum value; as K increases the scattering factor drops rapidly because of the factor K^2 in the denominator.

The calculation we have just given is valid only if the energy of the electron is much greater than the potential $V(\boldsymbol{r})$ inside the atom, since this is one of the conditions for use of the Born approximation. This assumption does not hold near the centres of heavy atoms and thus the scattering factors deduced for heavy atoms are liable to be inaccurate. For energetic electrons the electron mass m in Equ. 4.13 must be replaced by the relativistic expression $m = m_0(1 - v^2/c^2)^{1/2}$.

4.4 Neutron–nuclear scattering

The neutron wavelength is given by the relationship

$$\lambda = \frac{h}{\sqrt{2mE}} \quad \text{or} \quad E = \frac{h^2}{2m\lambda^2}$$

Thus the energy of a neutron whose wavelength is 1 Å is $\sim 1{\cdot}3 \times 10^{-13}$ erg so that it has translational energy equivalent to about 700 K.

There are two different interactions between a neutron and the atoms of a crystal which give rise to significant scattering; these are the neutron–nuclear interaction and the interaction between the neutron magnetic moment and any magnetic moment which the atoms of the crystal may possess. We will consider first the neutron–nuclear interaction which is just the same nucleon–nucleon force which is responsible for nuclear cohesion. As is well known, this is a strong interaction so that the potential of the neutron near the nucleus will be very different from its potential outside the crystal and the Born approximation should not be expected to hold. Fortunately, in order to predict the result of a neutron scattering experiment it is not necessary to have a precise model for forces which have a range much shorter than the neutron wavelength. We may use the method invented by Fermi and replace the true potential by a Fermi pseudo-potential. The pseudo-potential is chosen so as to give experimentally correct wavefunctions for the neutron when it is not in the near-neighbourhood of a nucleus. It is zero outside a radius r_0 from the nucleus and inside this radius has a constant value a; r_0 is of the order 10^{-13} cm. Use of the pseudo-potential rather than the true potential has an additional advantage: since it is obtained by smearing the potential out over a sphere of radius r_0, the pseudo-potential is small compared to the neutron energy and hence the Born approximation can be used to compute the scattering. In this approximation the wavefunction of elastically scattered neutrons can be written in the same form as Equ. 4.9 as

$$\psi_s(\boldsymbol{R}) = \frac{m}{2\pi\hbar^2} \frac{1}{|\boldsymbol{R}|} e^{i\boldsymbol{K}_0 \cdot \boldsymbol{R}} \int_{\text{crystal}} V(\boldsymbol{r})\, e^{i\boldsymbol{K} \cdot \boldsymbol{r}}\, d\tau_r$$

where m is now the mass of the neutron and $V(\boldsymbol{r})$ is the Fermi pseudo-potential. The atomic scattering factor for the neutron–nuclear interaction is therefore

$$f_n(\boldsymbol{K}) = \frac{m}{2\pi\hbar^2} \int_{\text{atom}} V(\boldsymbol{r})\, e^{i\boldsymbol{K}.\boldsymbol{r}}\, d\tau_r$$

$$= \frac{m}{2\pi\hbar^2} \left[\int_0^{r_0} a\, e^{i\boldsymbol{K}.\boldsymbol{r}}\, d\tau_r \right]$$

and since $r_0/\boldsymbol{K} \ll 1$

$$f_n(\boldsymbol{K}) = \frac{m}{2\pi\hbar^2}\, a(\tfrac{4}{3}\pi r_0^3) \tag{4.14}$$

Thus the atomic scattering factor is a constant independent of $\boldsymbol{K}$ and is usually denoted by b, the *nuclear scattering length.*

Since the nuclear scattering depends on the constitution of the scattering nucleus and not on the electronic configuration of the atom, the nuclear scattering length will be different for different isotopes of the same element. From the standpoint of nuclear scattering therefore, a crystal containing an element having more than one isotope will act as a randomly disordered system (see §3.12). The scattering in the Bragg reflections will be given by the mean scattering length which is known as the *coherent scattering length*; the incoherent scattering is proportional to the mean-square deviation of the scattering lengths, the *incoherent cross-section.* Additional incoherent scattering occurs when the scattering nucleus has non-zero spin because the neutron–nuclear interaction is spin dependent; thus the scattering length depends upon whether the neutron spin is parallel or anti-parallel to the nuclear spin. The spin dependence of the nuclear scattering is responsible for the high incoherent scattering cross-sections of hydrogen and vanadium.

4.5 Neutron–magnetic scattering

The magnetic moment in a crystal may have contributions from both electron spin and electron orbital angular momenta; both of these will contribute to the interaction with the neutron spin magnetic moment which gives rise to neutron–magnetic scattering. However in many cases, particularly for the important first series transition elements, the net orbital magnetic moment is small because of orbital quenching. We shall therefore consider only the more important spin contribution to the scattering. Because the neutron spin is small the scattering potential is weak, and hence the Born approximation is valid. In this case, however, the potential depends upon the relative orientations of the neutron spin and the electron spin, so the calculation of the scattered amplitude is more complicated and cannot be given here. The result for elastic scattering of unpolarised neutrons is that the differential scattering cross-section is given by

$$\frac{\partial\sigma}{\partial\Omega} = \left(\frac{2\gamma e^2}{mc^2}\right) (|P(\boldsymbol{K})|^2 - (P(\boldsymbol{K}) \cdot \hat{\boldsymbol{K}})^2) \tag{4.15}$$

here, m is the neutron mass and γ its magnetic moment in nuclear magnetons,

$$P(\boldsymbol{K}) = \int_{\text{crystal}} \psi^* \boldsymbol{S} \psi \, e^{i\boldsymbol{K} \cdot \boldsymbol{r}} \, d\tau_r \tag{4.16}$$

and $\boldsymbol{S}$ is the spin operator. Comparing Equ. 4.15 and 4.16 with Equ. 3.4, it can be seen that the single squared term of Equ. 3.4 is replaced by the sum of two similar squared terms in the neutron–magnetic scattering cross-section.

If we consider a simple ferromagnetic or antiferromagnetic crystal in which the spins are all parallel or antiparallel to a single magnetisation direction, then the function $P(\boldsymbol{K})$ can be written

$$P(\boldsymbol{K}) = \hat{\boldsymbol{\eta}} \int_{\text{crystal}} S(\boldsymbol{r}) \, e^{i\boldsymbol{K} \cdot \boldsymbol{r}} \, d\tau_r$$

$\hat{\boldsymbol{\eta}}$ is a unit vector parallel to the magnetisation direction and $S(\boldsymbol{r})$ is the spin density in the crystal which may be both positive and negative. The differential scattering cross-section then becomes

$$\frac{\partial\sigma}{\partial\Omega} = \left(\frac{2\gamma e^2}{mc^2}\right)^2 \sin^2 \alpha \left[\int_{\text{crystal}} S(\boldsymbol{r}) \, e^{i\boldsymbol{K} \cdot \boldsymbol{r}} \, d\tau_r\right]^2 \tag{4.17}$$

where α is the angle between $\hat{\boldsymbol{\eta}}$ and the scattering vector $\boldsymbol{K}$; $\sin^2 \alpha$ is zero (and hence the magnetic scattering is zero) when the magnetisation direction is parallel to the scattering vector and it is unity when the two are perpendicular. If we make the additional assumption that the spins are localised on the atoms and that each magnetic atom has a well-defined spin, then the term within square brackets in Equ. 4.17 can be expressed as

$$\sum_{j=1}^{n} S_j f_j(\boldsymbol{K}) \, e^{i\boldsymbol{K} \cdot \boldsymbol{r}_j} \tag{4.18}$$

S_j is the spin of the jth atom which has its centre distance $\boldsymbol{r}_j$ from the origin and

$$f_j(\boldsymbol{K}) = \frac{1}{S_j} \int_{\text{atom}} S(\boldsymbol{r})\, e^{i\boldsymbol{K}\cdot\boldsymbol{r}}\, d\tau_r$$

is the neutron–magnetic form factor. The magnetic structure factor is the summation of Equ. 4.18 taken over all the atoms in a magnetic unit cell.

It should be noted here that in many simple antiferromagnetic structures atoms having opposite spins are related to one another by one of the translational elements of the space group to which the crystal belongs (a lattice translation, a glide plane, or a screw axis). If this is the case, the magnetic structure factor will have finite values for the reflections which are normally systematically absent because of the element and they will be

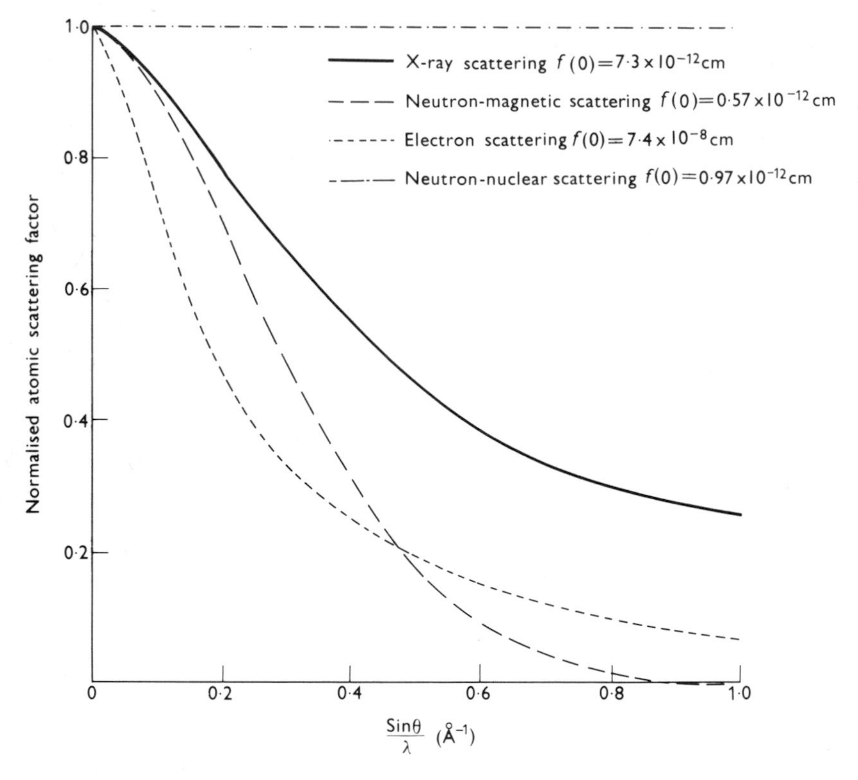

Figure 4.1 Normalised atomic scattering factors for iron.

systematically zero for the corresponding even-order reflections. The proof of this statement is left as an exercise for the reader. When both nuclear and magnetic scattering takes place with the same scattering vector, the resultant intensity for unpolarised neutrons is the sum of the nuclear and magnetic intensities. If the neutrons are polarised, interference between the nuclear and magnetic scattering can occur and the scattered intensity depends on the neutron polarisation direction. This effect will not be considered here.

The differences in the absolute magnitudes of the scattering amplitudes for X-rays, electrons and neutrons and their different degrees of variation with scattering angle are illustrated in Fig. 4.1.

4.6 Extinction

So far in this anlysis we have assumed that the scattering is so weak and the crystal size so small that there is no significant diminution in the intensity of the incident beam as it penetrates the crystal, either by true absorption or by the process of scattering. In real situations this is not a good assumption and these two processes must be taken into account. The extent to which the incident beam intensity is reduced because of the process of scattering clearly depends on the strength of the scattering interaction and is quite different for electrons, X-rays and neutrons. The importance of the effect can be judged by calculating the distance which the incident beam can travel in the crystal before this simple kinematical theory would predict that the whole of the incident intensity would be diffracted away. A simple calculation shows that this distance ($= \frac{1}{2}d_e$) is given approximately by

$$d_e = \frac{\pi V \cos\theta}{\lambda F(hkl)}$$

where V is the unit cell volume and d_e is known as the extinction distance.

For the (111) reflection from aluminium the values of d_e turn out to be 200 Å for 100 kV electrons, 7×10^{-4} cm for 1 Å wavelength X-rays and 6×10^{-3} cm for neutrons of the same wavelength. Thus the diminution of the incident beam cannot be neglected in electron diffraction except for *very* thin crystals, but the inaccuracy introduced by using the kinematical theory of X-ray and neutron diffraction is much less severe. Even so, the size of crystals commonly used in experiments is 10^{-2}–10^{-1} cm for X-rays and 0·2–2 cm for neutron diffraction, so that one might expect very large deviations from the kinematical theory. However, such large deviations are not observed except in the case of a few materials which grow as very perfect crystals. Normal crystals are made up of large numbers of small blocks of perfectly arranged crystal separated by regions containing imperfections, which may or may not give rise to small misorientations between the crystal blocks. If the linear dimensions of each crystal block are significantly less than d_e, it will scatter according to the kinematical theory. On passing from one block to another the coherence of the incident and scattered beams is lost, so that the total scattered intensity is the sum of the intensities scattered by each of the blocks individually. When the size of the blocks is too large for the kinematical approximation to be valid the crystal is said to exhibit *primary extinction.* The proper treatment of scattering from large perfect crystals requires consideration of the dynamic interaction between the incident and scattered beams and this is beyond the scope of this book. Treatments of the dynamical theory for X-ray and electron scattering may be found in James (1958) and Hirsch *et al.* (1965).

In cases in which the crystal blocks are small enough for the kinematical approximation to be valid, the total scattering may still be less than that predicted by the theory because of the process known as *secondary extinction.* Secondary extinction occurs when the misorientation of the crystal blocks is small enough, and the whole crystal large enough, for an appreciable part of the crystal well below the surface to have the same orientation as a part near the surface; the former then receives an incident beam which is reduced by the intensity scattered by the latter. It can be seen that secondary extinction will be greatest for those reflections with large structure factors for which the crystal orientation is such that the path length of the incident beam in the crystal is long. No completely satisfactory method has been developed for correcting measured integrated intensities for secondary extinction. Fortunately, in many crystals the effect is small except in the very strongest reflections.

5

Experimental Study of Diffraction by Crystals

5.1 The conditions for Bragg reflection

In Chapter 3 it was shown that the geometrical condition for a crystal to diffract radiation coherently with scattering angle 2θ is given by the Bragg equation, Equ. 3.12.

$$\lambda = 2d \sin \theta$$

d is the spacing of a set of planes in the crystal which are oriented such that their normal bisects the angle between the incident and diffracted beams. This orientation is also defined by the *Laue conditions*

$$2\pi \boldsymbol{K} \cdot \boldsymbol{a} = h\lambda; \quad 2\pi \boldsymbol{K} \cdot \boldsymbol{b} = k\lambda; \quad 2\pi \boldsymbol{K} \cdot \boldsymbol{c} = l\lambda \tag{5.1}$$

A single crystal placed in an arbitrary orientation in a monochromatic beam of radiation is rather unlikely to have any set of planes oriented so as to fulfil these conditions, so that diffraction will not usually occur. The techniques used for making diffraction measurements are ways of ensuring that during the experiment certain sets of planes within the crystal will fulfil the necessary conditions. There are three classes of technique which are commonly used. The moving-crystal technique includes all methods in which a single crystal is made to diffract radiation by moving it into the appropriate orientations. The Laue technique utilises polychromatic radiation and a stationary crystal: it relies on there being sets of planes in the crystal which are appropriately oriented to diffract some of the wavelengths within the primary beam. The third technique, the powder method, is that in which a polycrystalline rather than a single-crystal specimen is used; in this case a large number of crystal orientations is present in the specimen and the Bragg condition for diffraction of the monochromatic radiation by each set of planes should be fulfilled by some of the crystallites.

5.2 The Ewald construction

The Ewald construction is a very convenient method for representing the geometrical conditions for diffraction, using the reciprocal lattice representation. In Fig. 5.1, O is the origin of the reciprocal lattice and XC

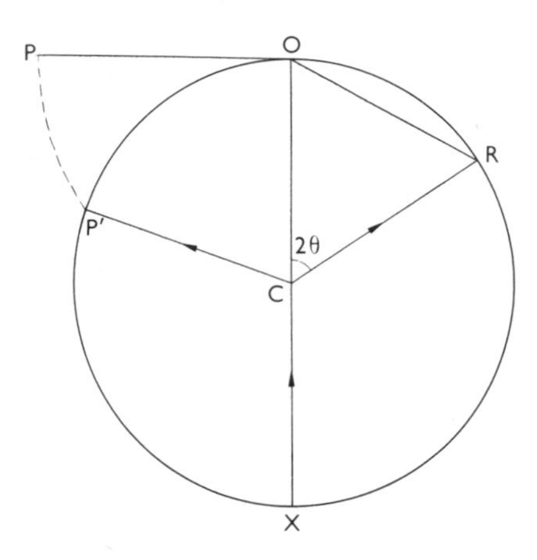

Figure 5.1 The Ewald construction.

is a unit vector in the direction of the incident radiation. R is a point anywhere on the surface of a sphere of unit radius centred on C. It can be seen that

$$OR = 2OC \sin \frac{O\hat{C}R}{2} \tag{5.2}$$

and if R is the reciprocal lattice point representing the set of planes hkl then $OR = \lambda/d_{hkl}$ and writing $O\hat{C}R = 2\theta$, Equ. 5.2 can be written

$$\lambda = 2d \sin \theta$$

from which it can be seen that the condition for Bragg reflection in the direction CR by the planes (hkl) is satisfied. Thus, in general, if P is a reciprocal lattice point, Bragg reflection from the planes represented by P will occur when the crystal is rotated so that P lies on the surface of the sphere at P′ and the direction of the diffracted radiation will be given by CP′.

5.3 Diffractometer geometries

Perhaps one of the most straightforward and easily understood ways of recording the radiation diffracted by a crystal is by using a diffractometer: this method is applicable to both X-ray and neutron diffraction. The ionisation spectrometer used by the Braggs to determine the first crystal structures was a simple diffractometer, and although great improvements have taken place in the intervening years in methods of detecting radiations present day diffractometers are essentially similar. The essential parts of such an instrument are a collimating system for the incoming radiation, a means for orienting the crystal, and a radiation detector (counter) which can be rotated so as to point at the correct angle to accept the diffracted beam. Three commonly used diffractometer geometries are illustrated in Fig. 5.2, they differ only in the relative orientations of the crystal rotation axis and the incident and diffracted beams.

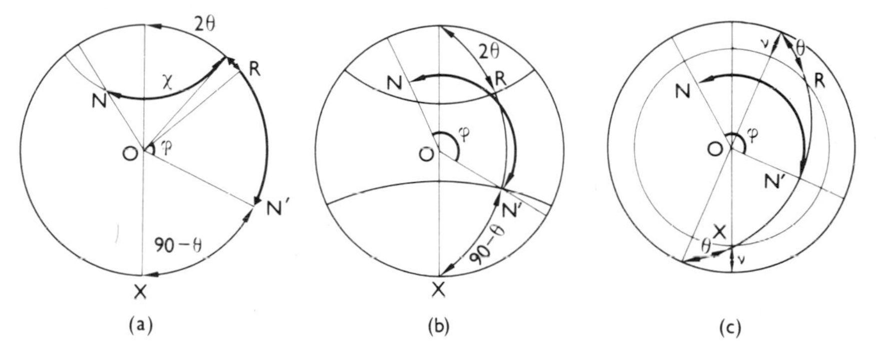

Figure 5.2 Stereograms illustrating the orientations of the normal to the diffracting planes N and the directions of the incident and diffracted rays X and R for three different geometries (a) the equatorial method, (b) the normal-beam method and (c) the equi-inclination method.

In the *equatorial* method, (a), this rotation axis is always perpendicular to both incident and diffracted beams; the crystal orienter must permit rotation about this axis and at least one other, represented by X in the stereogram. In the case illustrated, the required rotations to bring the normal N into its reflecting position N′ are χ about X followed by ϕ about O. Often the crystal orienter will have a further degree of freedom allowing greater flexibility in the orientations of the other two circles, which may thus be positioned so as not to interfere with the incident and diffracted beams.

In both the *normal-beam* and *equi-inclination* methods, (b) and (c), the crystal is oriented so that a prominent zone-axis $[UVW]$ is parallel to the rotation axis O. Reflections in the $[UVW]$ zone have their normals in the equatorial plane, and for these reflections all three methods are equivalent. For a general reflection hkl, however, the normal does not lie in the equatorial plane but the reciprocal lattice points representing the set of planes for which $hU + kV + lW = n$ lie on a plane distance

$$\zeta = \lambda n / |U\boldsymbol{a} + V\boldsymbol{b} + W\boldsymbol{c}|$$

from the origin. The diffracted rays from this set of planes lie on the surface of a cone, of semi-angle $(90 - \nu)$ with $\nu = \sin^{-1} \zeta$, whose apex is at the centre of the Ewald sphere. In the normal-beam method, the crystal rotation axis O is kept normal to the incident beam. In order to accept the diffracted beams, the counter must be inclined at an angle $(90 - \nu)$ to O as well as being rotated by the appropriate angle about O. In the equi-inclination method, on the other hand, both the incoming X-ray beam and the counter are tilted at an angle $(90 - \frac{1}{2} \sin^{-1} \zeta)$ to O. The crystal rotation axis is thus equally inclined to both incident and diffracted beams.

5.4 Moving-crystal methods using photographic recording

In X-ray diffraction the quantum detector of the diffractometer can be replaced by a photographic emulsion. This is usually wrapped around the axis of rotation of the crystal to form a cylinder. In the most commonly used arrangement, the X-rays enter between the two ends of the film and the undiffracted beam passes out through a small hole punched in the middle.

If the crystal is rotated by 360° about its axis, all the reciprocal lattice points within the toroidal volume illustrated in Fig. 5.3 will pass through the surface of the Ewald sphere. Hence, the planes which they represent will have an opportunity to diffract the incident beam.

Latent images are formed at the points at which the diffracted beams strike the film and on development a picture such as that illustrated in Fig. 5.4a is obtained. The origin of the horizontal lines of spots known as '*layer lines*' is clear from the considerations of the previous section. If the crystal is rotated about a zone axis $[UVW]$, successive layers correspond to reflections from planes (hkl) for which $hU + kV + lW = 0, 1, 2, \ldots n$. The common azimuthal angle of beams diffracted by planes belonging to the nth layer is $(90 - \nu_n)$ where

$$\nu_n = \sin^{-1} \frac{n\lambda}{|U\boldsymbol{a} + V\boldsymbol{b} + W\boldsymbol{c}|} = \sin^{-1} \zeta_n \tag{5.3}$$

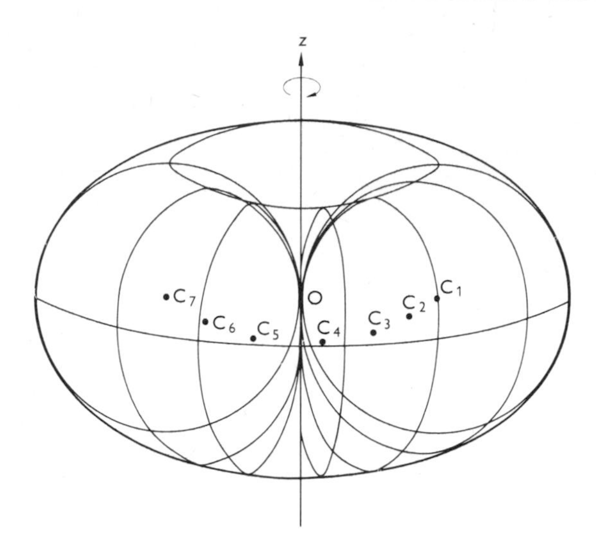

Figure 5.3 The volume of reciprocal space swept out by the Ewald sphere when a crystal is rotated about an axis z.

If the camera radius is r and $2S_n$ is the distance on the photograph between the nth layers above and below the equatorial plane then

$$\nu_n = \tan^{-1}\frac{S_n}{r} \quad \text{or} \quad Ua + Vb + Wc = \frac{1}{n\lambda}\sin\left(\tan^{-1}\frac{S_n}{r}\right) \tag{5.4}$$

Thus the distance between the layer lines on the rotation photograph gives the distance between lattice points in the crystal along the zone axis direction which is parallel to the axis of rotation.

Unless the unit cell of the crystal is small, or rather symmetrical, it is not usually possible to assign a unique set of indices to the individual reflections on a rotation photograph, particularly if the cell dimensions are not already known with some accuracy. The number of reflections which occur on a single photograph may be substantially reduced by oscillating the crystal through a small angle—usually 5, 10 or 15 degrees—rather than rotating it freely, (Fig. 5.4b). In this case only planes represented by reciprocal lattice points which pass through the surface of the reflecting sphere during the oscillation have a chance to diffract. The method used for indexing the reflections is illustrated for two layers in Fig. 5.5.

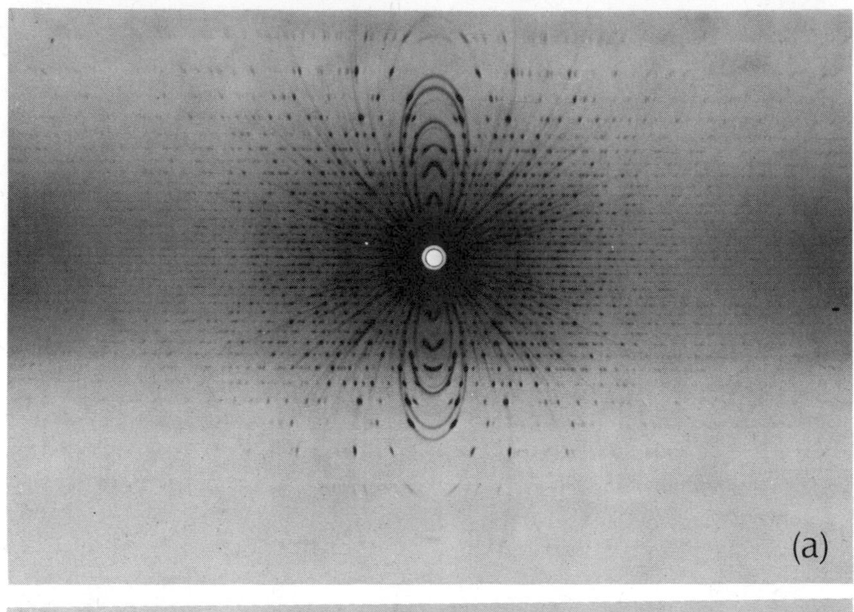

Figure 5.4 (a) A rotation photograph of the intermetallic compound VAl_{10} taken with Mo$K\alpha$ radiation. One of the axes of the cubic unit cell is parallel to the axis of rotation. (b) A 5° oscillation photograph of $FeCO_3$ taken with Cu$K\alpha$ radiation $\lambda = 1{\cdot}54$ Å. One of the axes of the rhombohedral unit cell is parallel to the axis of oscillation.

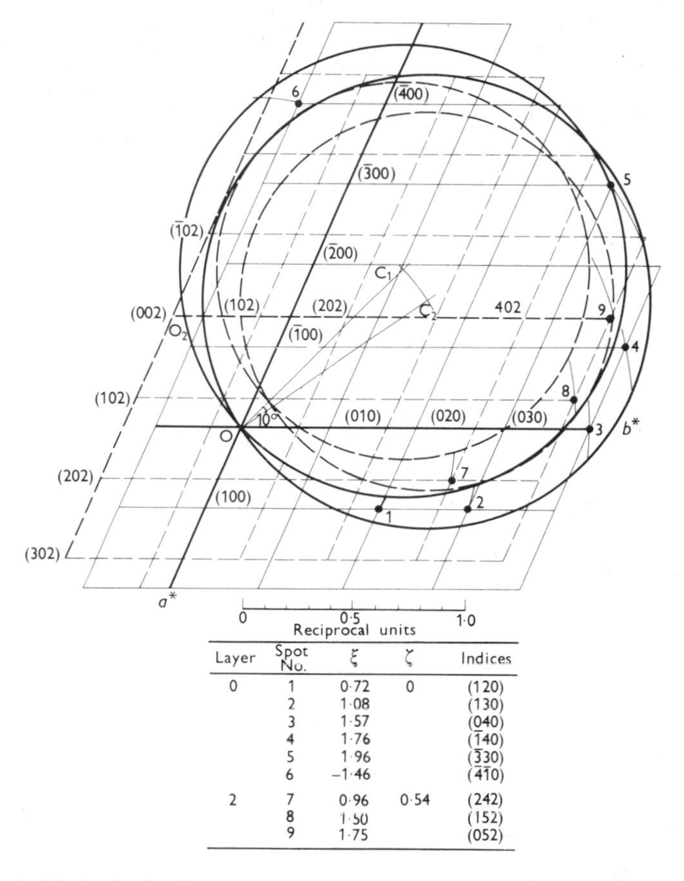

Layer	Spot No.	ξ	ζ	Indices
0	1	0·72	0	(120)
	2	1·08		(130)
	3	1·57		(040)
	4	1·76		$(\bar{1}40)$
	5	1·96		$(\bar{3}30)$
	6	−1·46		$(\bar{4}\bar{1}0)$
2	7	0·96	0·54	(242)
	8	1·50		(152)
	9	1·75		(052)

Figure 5.5 Method of indexing the zero and second layers of the photograph shown in Fig. 5.4b.

$a^* = b^* = c^* = 0{\cdot}3928$; $\lambda = 1{\cdot}542$ Å;

$\alpha^* = \beta^* = \gamma^* = 113°44'$

The full lines indicate the zero layer of the reciprocal lattice and the full circles the extreme sections of the reflecting sphere. The broken lines and circles refer to the second layer.

The horizontal deflection Y of the incident beam is given by t/r where t is the distance to the spot measured from the central line on the film and

$$\sin\frac{Y}{2} = \frac{\xi}{2}$$

where ξ is the projection of the relevant reciprocal lattice vector on to the equatorial plane. Values of ξ and ζ for each reflection on the film may be deduced from measurements on the film, or alternatively their values may be obtained directly using a Bernal chart. Bernal charts which correspond to all the commonly used camera radii are obtainable and on them lines of constant ξ and constant ζ are drawn at equal intervals, usually 0·05. The charts, which are usually drawn on transparent paper, are placed over the film so that the centres and equatorial planes coincide and this enables the values of ξ and ζ for each reflection to be read. A scale drawing of each layer of the reciprocal lattice is made and on it circles corresponding to the extreme positions of the reflecting sphere during the oscillation are drawn. The reflections seen on the photograph should then lie in the area between these circles. Arcs of radius ξ corresponding to each reflection in the layer are drawn (also to scale) and the intersections of these arcs with reciprocal lattice points inside the relevant area enable the reflections to be indexed. In the example illustrated (Fig. 5.5), it should be noticed that the zone axis is not parallel to a reciprocal lattice vector and so the reciprocal lattice points in different layers do not lie directly over one another.

Interpretation of oscillation photographs of substances whose unit cells are completely unknown is much more complicated. It is usually simpler to use a moving-film technique to obtain more information about the relative positions of reciprocal lattice points within a layer. The discussion of such methods is beyond the scope of this book, but further information about such techniques is given in Henry, Lipson and Wooster (1961) and Buerger (1960).

5.5 The Laue method

The Laue method is perhaps the simplest of all techniques for obtaining diffraction photographs. It was using this method that Max von Laue, in 1912, obtained the first X-ray diffraction photograph thus demonstrating conclusively the wave nature of X-radiation.

In the Laue method a single crystal is mounted so that a collimated beam of white radiation can fall upon it and a film placed so that the diffracted radiation can be recorded. If the crystal is thin enough for the radiation to pass through it the film may be placed behind the specimen. Alternatively, the film may be placed between the radiation source and the specimen and the incident beam allowed to irradiate the crystal through a hole cut in the centre of the film. This is known as the back-reflection method. The number of reflections which can be recorded on a single flat flim is rather limited; a greater number are obtained by using a cylindrical film wrapped around the specimen. Examples of these three types of Laue photograph are shown in Fig. 5.6.

It can be seen that the spots on these photographs are clustered along well-defined curves. The origin of these curves may readily be understood by means of the reciprocal lattice and Ewald sphere. The reciprocal lattice points, which represent a single set of planes for the range of wavelengths in the incident beam, are distributed along the radius vector in the reciprocal lattice which is normal to that set of planes. The point on this radius vector which is unit distance from the centre of the Ewald sphere will give rise to a diffracted beam. Now the sets of planes in a crystal may be grouped into zones, all planes in a zone having their normals in a common plane perpendicular to the zone axis. The locus of the diffracted beams from the set of planes forming a zone is the cone formed by lines joining the centre of the sphere to that circle in which the plane through the origin of the reciprocal lattice, perpendicular to the zone axis, intersects the sphere. Thus the axis of the cone of diffracted rays is parallel to the zone axis and the direction of the incident beam lies in its surface. The intersections of these cones with the film give the characteristic curves on the Laue photographs, each of which therefore corresponds to a zone of reflections. The formation of these curves for each of the three common geometries is shown in Fig. 5.7.

The Laue method is used extensively for determining the orientation of crystals and for establishing their symmetries. It cannot easily be used in the determination of lattice constants or integrated intensities, for which a knowledge of the diffracted wavelength is essential. The photographs of Fig. 5.6 illustrate how the symmetry of a crystal can be revealed by Laue photographs: (a) shows a hexagonal axis aligned parallel to the X-ray beam and (c) shows the intersection of two mirror planes. It must be realised that the crystal symmetry cannot be determined uniquely from Laue photographs because of Friedel's law (§3.11) and the photograph (c) could equally well correspond to the intersection of two diad axes, or to a diad axis lying in a mirror plane.

5.6 Determination of crystal orientation by the back-reflection Laue method

The back-reflection Laue method is one of the most useful techniques for the determination of the orientation of large single crystals. Since such a determination of orientation is a necessary preliminary to many experiments in solid-state physics, it is worth going into in rather more detail. The camera used is very simple consisting of a plateholder in the centre of which the collimator, through which the incident beam passes, is fixed.

(a)

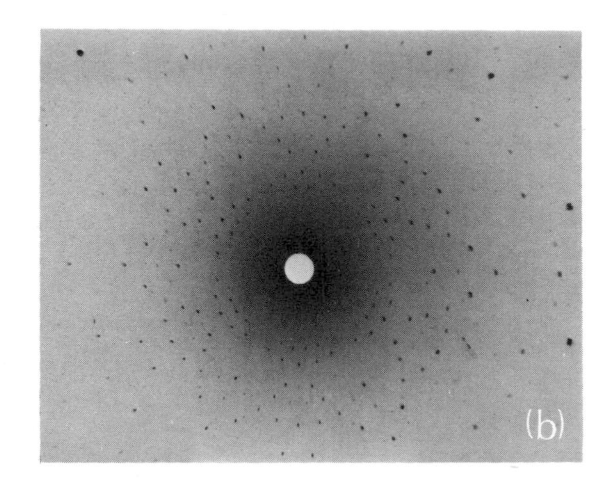
(b)

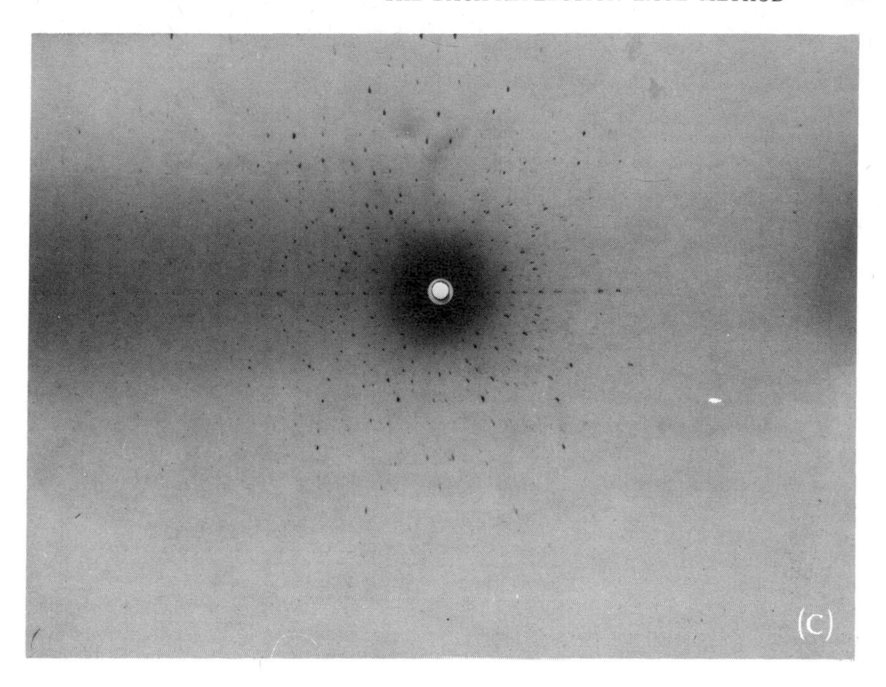

Figure 5.6 Laue photographs (a) Back-reflection photograph taken with a polaroid camera and fluorescent screen. (b) Forward-reflection photograph on a flat plate showing a hexagonal axis parallel to the X-ray beam. (c) Photograph taken using a film in a cylindrical cassette shows mirror planes parallel and perpendicular to the X-ray beam.

The crystal is then supported on some suitable mount so that the collimated beam hits it at a point a known distance from the film plate. Figure 5.6b is an example of a photograph of diamond taken by this method. The intersections of the cones of diffracted rays from each zone with the plane of the film are hyperbolae. It can readily be seen that the shape of these hyperbolae, though not necessarily the distribution of reflections on them, must be symmetrical about the line of intersection of the plane, which contains the incident beam direction and the zone axis, with the plane of the film.

Referring to Fig. 5.8a, which illustrates the interpretation of a hypothetical photograph showing three zones, the line of symmetry for zone 1 is OA and it intersects the hyperbola at A, a distance x_1 from the centre

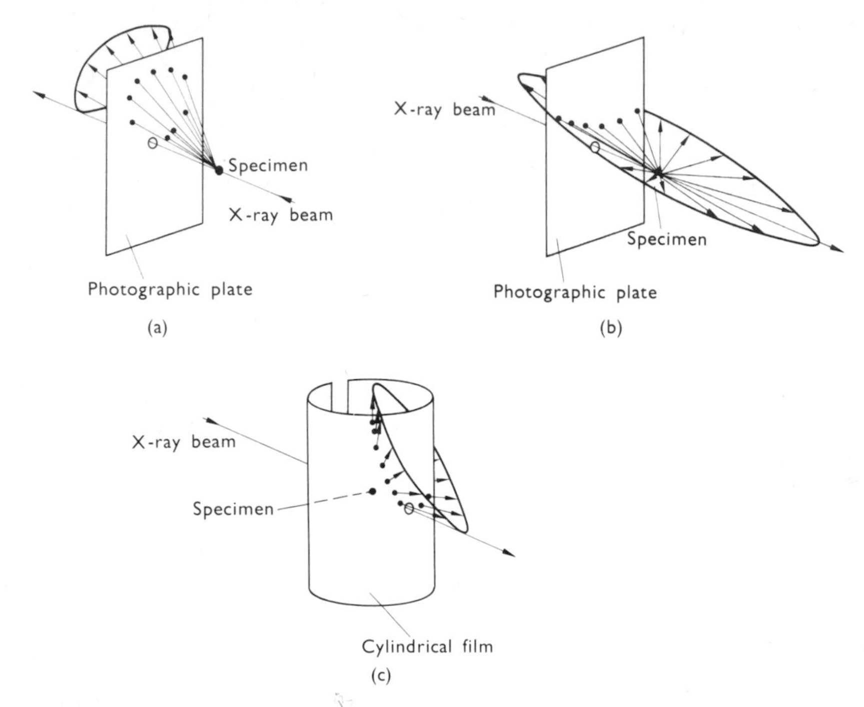

Figure 5.7 Diagrams showing the formation of the curves given by zones of reflections in three different Laue geometries. (a) Forward reflection; (b) back reflection; (c) cylindrical geometry.

of the film. If ϕ_1 is the angle between the zone axis and the incident beam then

$$\frac{x_1}{r} = \cot \phi_1$$

where r is the specimen-to-film distance.

The pole of the zone axis may readily be plotted on a stereographic projection as illustrated in Fig. 5.8b. To determine the orientation of the crystal completely, as many zones as possible are plotted on the stereographic projection and these are then indexed by comparing the angles between them with the calculated interzonal angles. For the cubic system the principal interzonal angles are tabulated in the *International Tables for Crystallography*, Volume II, (1959), p. 120. For other systems the angles depend upon the axial ratios and interaxial angles; they are usually obtained most easily by measurement of a stereographic projection of the crystal drawn in the standard orientation.

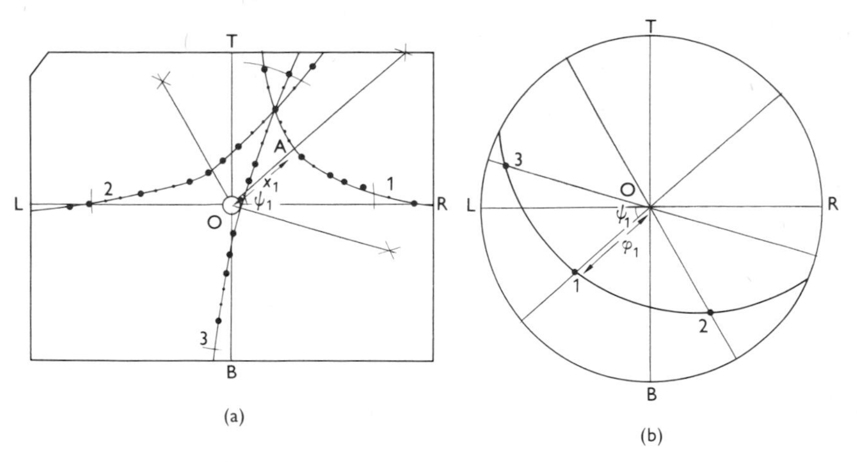

Figure 5.8 Diagram illustrating the interpretation of a back-reflection Laue photograph. This method is commonly used for determination of the orientation of large single crystals.

5.7 The powder method

For the powder method the specimen must be ground to a fine powder (about 250/300 mesh size) and mounted in such a way that it can be bathed in a monochromatic X-ray beam. In the simplest method for obtaining powder photographs the specimen is shaped into the form of a cylinder, of the order of 0·5 mm diameter, either by packing the powder into a fine Lindemann glass capillary or by using a binder such as Gum Tragacanth. The specimen is mounted at the centre of the camera and is usually rotated about its own axis. A narrow strip of film is pressed against the inside of a cylindrical holder, the axis of which coincides with the axis of the specimen: it has a hole or holes punched in it to allow the entry and/or exit of the primary beam. The three commonly used film mountings are shown in Fig. 5.9; of these the Straumanis, or asymmetric film mounting (c), is the most generally useful.

Figure 5.10 shows some examples of powder photographs; the X-ray reflections may be seen to occur as 'lines' on the photographs. Each of these lines corresponds to the reflections from all planes of one form (see §1.5). This occurs because the crystal is a fine powder and, if the orientation of grains within the powder may be supposed to be random, there will be some particles oriented so as to diffract radiation at the Bragg angle for every possible set of planes in the crystal. Consider a set of planes with spacing d, the spacing of all other planes of the same form will also be d and these planes will diffract radiation of wavelength λ at an angle θ such that

$$\sin \theta = \frac{\lambda}{2d}$$

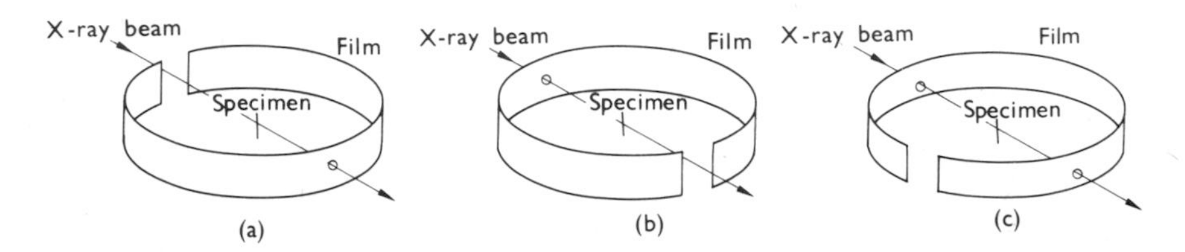

Figure 5.9 The relative orientations of the film and incident X-ray beam in the three most common methods used for powder photography.

Since all orientations of the crystallites should be equally likely, the rays diffracted by these planes will lie along the surface of a cone of half-angle 2θ whose axis is the incident beam. The 'lines' on the powder photographs are the intersections of this cone with the cylindrical strip of film. Figure 5.11 illustrates the interpretation of a photograph of a cubic material taken with the asymmetric film mounting. Table 5.1 shows the steps in the analysis of the photograph of Figure 5.10a.

For the cubic system

$$\sin^2 \theta = \frac{\lambda^2}{4a^2}(h^2 + k^2 + l^2) \tag{5.5}$$

so that the values of $\sin^2 \theta$ are proportional to integers which are the sum of squares of three integers. All integers less than 20, except for 7 and 15, can be represented in this way. Inspection of the values of $\sin^2 \theta$ usually enables gaps, corresponding to 7 or 15, to be located and hence the reflection indices to be deduced. Alternatively, values of $\sin^2 \theta$ may be plotted on a strip of paper and matched against a set of straight lines representing the variation of $\sin^2 \theta$ with $\lambda^2/4a^2$ for different sets of hkl (Fig. 5.12).

The indexing of powder photographs of non-cubic materials is straightforward, so long as the dimensions of the unit cell are known. If the cell is not known, there is no simple way of indexing the lines directly. The most generally useful methods are based on that due to Ito (1950). This is a trial and error method in which three low angle lines are arbitrarily chosen

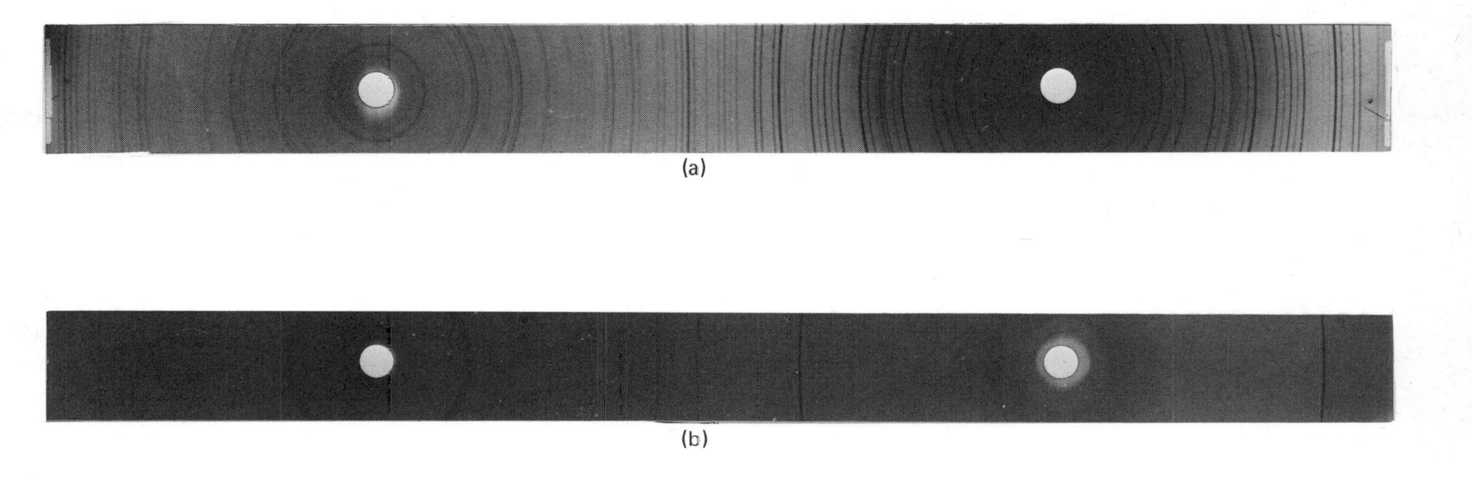

(a)

(b)

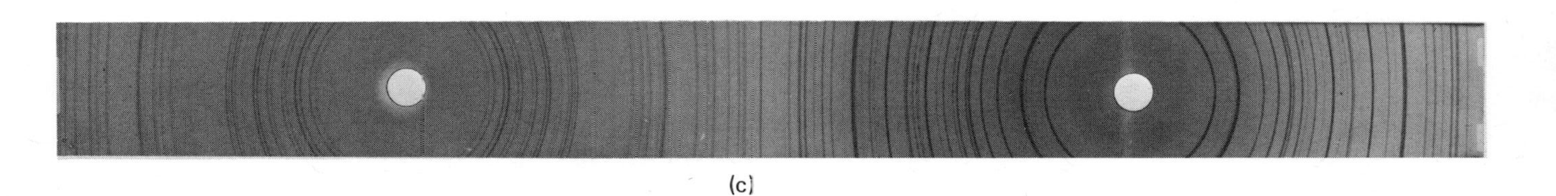

(c)

Figure 5.10 Three typical powder photographs. (a) $NaClO_3$, a cubic material with a fairly large unit cell, taken with Cu$K\alpha$ radiation $\lambda = 1{\cdot}54$ Å. (b) Iron taken with Cr$K\alpha$ radiation $\lambda = 2{\cdot}2$ Å. (c) Quartz taken with Cu$K\alpha$ radiation.

as (100), (010) and (001). Pairs of lines corresponding to (101) and (10$\bar{1}$) etc. are then sought and these will of course be coincident if the appropriate interaxial angle is 90°. A method such as this is very suitable for use with a computer which can try a large number of different trial solutions and select that giving the best fit with the largest number of reflections. The unit cell given by such a method will not necessarily be the conventional one, exhibiting the symmetry of the crystal, and this has therefore to be deduced subsequently.

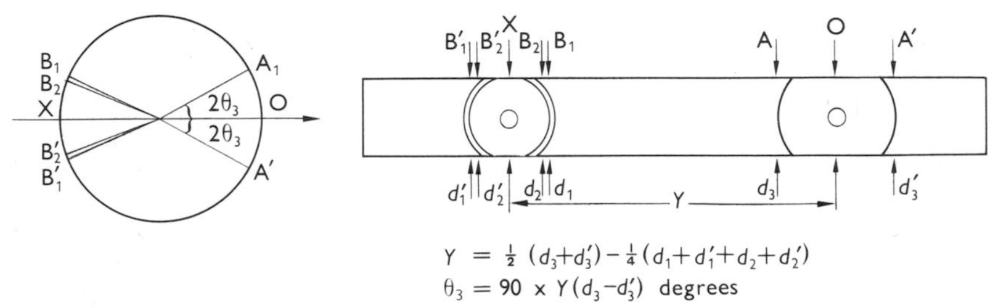

Figure 5.11 The calculation of Bragg angles from powder photographs taken with the asymmetric film mounting.

Powder photographs of low symmetry materials or of those with large unit cells often show a very large number of lines which, except at the lowest angles, may not be properly resolved from one another. It is almost impossible to index the lines in such photographs uniquely, unless the lattice parameters are already known with high accuracy. In addition, because of overlap, the intensities of the lines cannot be measured accurately. For this reason, quantitative measurements of the intensities scattered by materials of this sort cannot be made on powders and recourse has to be made to single-crystal methods.

5.8 Use of powder photographs to identify materials

One use that can be made of the distinct, though often complex, powder patterns given by different materials is as a means of identification. The American Society for Testing Materials (A.S.T.M.) has developed an index in which the *d*-spacings and relative intensities of the lines in the powder photographs of a large number of materials are listed. Each distinct crystalline phase occurring in the index is represented by three cards. On one of these all the *d*-spacings and intensities are tabulated and this is filed in the index in a position corresponding to the *d*-spacing of the strongest line in

Table 5.1 The analysis of the powder photograph shown in Fig. 5.10a to deduce the lattice parameter for $NaClO_3$. The upper half of the table is for measurements about the exit hole, O. The remainder are measurements about to entry hole, X (See also Figs. 5.11, 5.12 and 5.13.)

d_1 cm	d_2 cm	(d_1+d_2)	(d_1-d_2)	$\theta°$	$\sin\theta$	$\sin^2\theta$	N	hkl
25·710	21·904	47·614	3·806	9·544	0·16580	0·0275	2	(110)
26·142	21·468	47·610	4·674	11·720	0·20313	0·0413	3	(111)
26·510	21·102	47·612	5·408	13·561	0·23448	0·0550	4	(200)
26·836	20·774	47·610	6·062	15·201	0·26221	0·0688	5	(210)
27·136	20·474	47·612	6·660	16·701	0·28738	0·0826	6	(211)
27·678	19·952	47·630	7·726	19·374	0·33174	0·1100	8	(220)
27·914	19·700	47·614	8·214	20·597	0·35179	0·1238	9	(221)(300)
28·152	19·470	47·622	8·682	21·771	0·37090	0·1376	10	(310)
	mean	47·6142						
	$\frac{1}{2}$ mean	23·8071	$(\frac{1}{2}\text{mean} - d_2)$					
	16·318		7·489	37·559	0·60958		27	(511)(333)
	15·998		7·409	39·164	0·63154		29	(520)(432)
	15·838		7·969	39·966	0·64233		30	(521)

d_1	d_2	(d_1+d_2)	(d_2-d_1)	$90-\theta$	$\sin\theta$	$\frac{\lambda}{2\sin\theta}$	N	$\sqrt{N}$	a	Nelson-Riley fn.
2·040	9·684	11·724	7·644	19·168	0·944652	0·815412	65	8·06225	6·5741	0·101
2·118	9·600	11·718	7·482	18·762	0·946885	0·815515	65		6·5749	0·097
2·298	9·424	11·722	7·126	17·869	0·951858	0·809239	66	8·12404	6·5743	0·087
2·390	9·336	11·726	6·946	17·418	0·954230	0·809237	66		6·5743	0·082
2·874	8·852	11·726	5·978	14·990	0·965971	0·797416	68	8·24621	6·5757	0·060
2·982	8·738	11·720	5·756	14·434	0·968455	0·797351	68		6·5751	0·055
3·206	8·520	11·726	5·314	13·325	0·973120	0·791558	69	8·30662	6·5752	0·047
3·332	8·394	11·726	5·062	12·693	0·975584	0·791524	69		6·5749	0·043
3·578	8·144	11·722	4·566	11·450	0·980098	0·785922	70	8·36660	6·5755	0·034
3·712	8·002	11·714	4·290	10·758	0·982476	0·785972	70		6·5759	0·031
4·608	7·114	11·722	2·506	6·284	0·994007	0·774925	72	8·48528	6·5754	0·010
4·892	6·832	11·724	1·940	4·865	0·996412	0·774979	72		6·5759	0·006
	mean	11·723								
	$\frac{1}{2}$ mean	5·8615								

Difference between means = 17·9456 cm = 180° of 2θ.

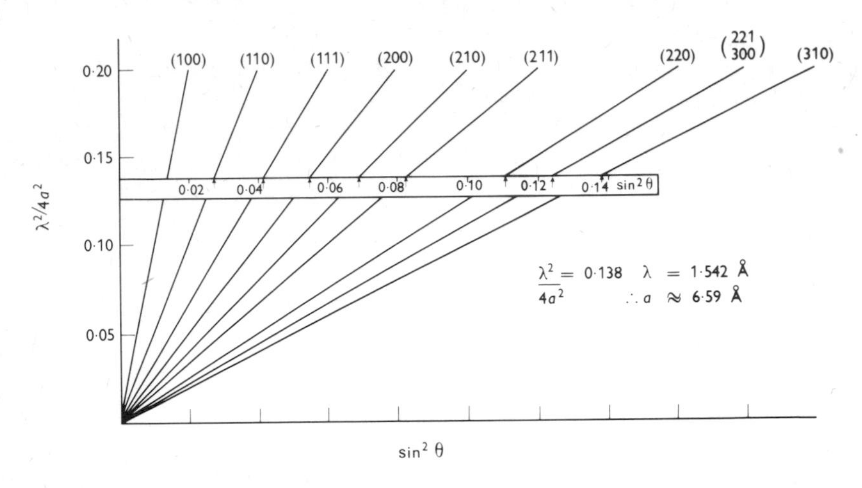

Figure 5.12 A method of indexing the powder photograph shown in Fig. 5.10a.

the pattern. The other two cards have on them only the name of the phase and information about the three strongest lines: they are filed in positions corresponding to the *d*-spacings of the second and third strongest lines. This system makes it possible to identify an unknown material by calculating the *d*-spacings of the lines in a powder photograph and selecting the three most intense ones. That part of the index corresponding to *d*-spacings near to that of the strongest line is searched for a card for which the other *d*-spacings and the relative intensities of the lines are also in agreement. The tabulation under second and third strongest lines, in addition to the strongest, enables impurities or mixtures to be identified even though the strongest line may be overlapped by that of another constituent.

5.9 Determination of accurate lattice parameters

Examination of the photographs of Fig. 5.10 will show that around the left-hand hole in the film each of the lines is a doublet. The reason for this lies in the nature of the Bragg equation, differentiation of which gives

$$d\lambda = \frac{1}{2d} \sec\theta \, d\theta$$

$$\frac{d\lambda}{\lambda} = \cot\theta \, d\theta$$

$$d\theta = \tan\theta \frac{d\lambda}{\lambda} \quad \text{or} \quad \tan\theta \frac{d(d)}{d} \tag{5.6}$$

Thus the change in angle brought about by a small change in either the spacing or the wavelength depends upon tan θ and therefore tends to infinity as θ tends to 90°. The Kα X-ray line, which is the strongest emission line from most targets and hence the one most commonly used, is a doublet with a wavelength separation for copper of 0·0038 Å; it is resolved at high angles and the doublet lines may be seen at the high-angle ends of the photographs of Fig. 5.10. The good resolution obtained in the high-angle region of a powder photograph enables very accurate values of the lattice parameters to be obtained. With $\theta = 85°$ a change in d-spacing of 0·001 Å causes a shift in angle of $\frac{1}{2}°$, which is easily measurable.

Looking at the possibilities from a different point of view, it should be possible with care to determine the position of the centre of a powder line to 0·2 mm. If the camera radius is 5·7 cm, this corresponds to an angular accuracy of 0·1° or a sensitivity of 0·0002 Å at $\theta = 85°$. In order to achieve an accuracy in lattice parameter determination of this order, various effects which lead to systematic errors in measurement of the Bragg angle must be eliminated or corrected. We have seen that as θ approaches 90° the error in d caused by a small error in θ becomes vanishingly small. These systematic effects would therefore not cause any error in the lattice parameters were it possible to record reflections with $\theta = 90°$. In practice, the geometry of the method restricts the highest angle to about 87·5° because the angle subtended by the collimator at the specimen cannot be made less than about 10°. It is possible however to deduce the lattice parameter which would have been obtained from a measurement at 90° by extrapolating the values obtained at lower angles against some suitable function of θ. The extrapolated result at $\theta = 90°$ will be most accurate if the function of θ used is one which can be expected to give a linear extrapolation. Hence it is necessary to consider which are the principal sources of error and how they depend upon θ in any particular case. These errors may be due to mis-centering of the specimen in the camera, divergence of the X-ray beam, absorption in the specimen and film shrinkage. Other less important errors are caused by the finite height of the specimen and refraction of the X-ray beam. The effects of uniform film shrinkage are eliminated by use of the asymmetric film mounting, which also removes the need for the camera radius to be known accurately. For most powder specimens the main angle-dependent error is that due to absorption of the X-ray beam in the specimen, which causes a shift in the position of the peak of the powder line. It has been shown that in these circumstances the error in the deduced lattice parameter is approximately proportional to $\frac{1}{2}(\cos^2\theta/\theta + \cos^2\theta/\sin\theta)$, the Nelson–Riley function. In the case of a non-absorbing specimen the principal error is probably that due to mis-centering, and the appropriate extrapolation function is then $\cos^2\theta$.

Figure 5.13 shows the result of extrapolating the high-angle data of Table 5.1 against $\frac{1}{2}(\cos^2\theta/\theta + \cos^2\theta/\sin\theta)$. It can be seen that the result is very strongly influenced by the highest angle reflection and that the influence of reflections with θ less than 75° is rather small. If good values for

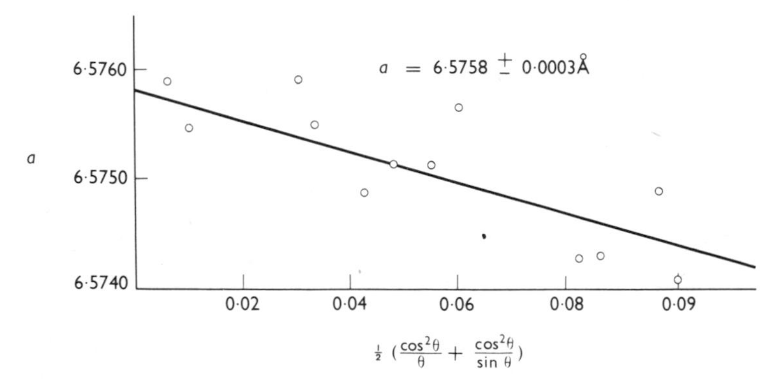

Figure 5.13 The determination of accurate lattice parameters by extrapolation to $\theta = 90°$.

lattice parameters are to be obtained, it is very important to choose a radiation which will give at least one reflection with θ greater than 80°.

The procedure to be followed for obtaining the lattice parameter of cubic crystals will be clear from Section 5.7. For non-cubic crystals the analytical least-squares method due to Cohen (1935) is probably the most useful, particularly as the calculations may easily be programmed for a computer. For any particular crystal system, the spacing d of planes (hkl) may be written in terms of the lattice constants (§1.10). For instance, in the orthorhombic system

$$\frac{1}{d^2} = \frac{h^2}{a^2} + \frac{k^2}{b^2} + \frac{l^2}{c^2}$$

Suppose the extrapolation function appropriate to d is $f(\theta)$, then the spacing d can be written in terms of the observed spacing d_0 by an equation such as

$$d = d_0(1 - \mathrm{k}f(\theta)) \tag{5.7}$$

where k is a small constant which is the same for all measurements from one photograph. Then

$$\frac{1}{d^2} \simeq \frac{1}{d_0^2}(1 + 2\mathrm{k}f(\theta))$$

so that

$$\frac{h^2}{a^2} + \frac{k^2}{b^2} + \frac{l^2}{c^2} - \frac{4 \sin^2 \theta}{\lambda^2} = \frac{8}{\lambda^2} \mathrm{k}f(\theta) \sin^2 \theta \qquad (5.8)$$

This equation has four adjustable parameters and the set of equations corresponding to a set of measurements may be solved by the usual least-squares techniques for a, b, c and k. An alternative graphical method is to divide the observed reflections into groups: those with lc^* greater than $ha^* + kb^*$, those with kb^* greater than $ha^* + lc^*$ and those in which ha^* is greater than $kb^* + lc^*$. Using approximate values for a and b, the first group of reflections may be used to obtain an accurate value for c. Similarly the second and third groups may be used to obtain better values for b and a respectively. The calculation is then repeated but using the better values of a, b and c and the process re-cycled until no further significant changes occur. In order to obtain high-angle reflections in all three groups, it may be necessary to take photographs with X-rays of several different wavelengths.

5.10 The powder diffractometer

For many applications the powder diffractometer has nowadays replaced the powder photograph for recording diffraction patterns from polycrystalline samples though it cannot easily be applied when only very small quantities of material are available ($< 0{\cdot}05$ cm^3). The theory behind the method is unchanged. The specimen is usually pressed into the form of a flat plate of sufficient thickness to absorb the whole incident beam. The plane of the specimen is set so that the normal to it bisects the angle between the incident beam and the centre line of the X-ray detector. This detector is a quantum counter, usually a Geiger, proportional or scintillation counter and ideally records one count in a scaler for each X-ray quantum incident on it. The shaft which carries the detector can be rotated about the axis of the specimen and can be set at an angle 2θ to the direction of the incident beam. The detector is aligned so that its centre line always passes through the axis of rotation of the specimen in a direction normal to it. The X-ray intensity scattered by the sample is recorded as a function of θ, either by carrying out a continuous scan or using a step-scanning method. In the continuous scan, the output of the detector is fed into a rate-meter and chart recorder. In the step-scanning method, the number of quanta detected in a fixed time is recorded with the specimen and detector stationary and between each recording they are stepped on by a small angular distance. The diffractometer method has the advantage that it is relatively easier to obtain quantitative estimates of the intensities of the diffraction lines and this is of considerable value in estimating the relative proportions of phases in a mixture. Measurements are obtained more rapidly both because the sensitivity of quantum detectors is high and because the geometry of the method enables a larger volume of sample to be irradiated.

Neutron-diffraction data from polycrystalline samples are almost always collected using a diffractometer method as there is as yet no satisfactory neutron-sensitive film having a low enough noise level to be useful. The sample in the neutron powder diffractometer is usually held in a thin-walled cylindrical aluminium can, 1–2 cm in diameter. It is not usually practicable to use a plane sample because of the low absorption of thermal neutrons in most materials. One useful modification which can be made to the conventional diffractometer is to mount several detectors on the detector arm, rather than one. As the arm moves round, each of the detectors records a different part of the pattern and hence the whole pattern can be recorded in a much shorter time than if just a single detector is used.

In Fig. 5.14, the X-ray and neutron powder patterns of NiO are compared. Features to note are the extra lines on the neutron pattern given by the antiferromagnetic structure of NiO, which has a repeat distance twice that of the chemical structure, and the difference in width of the lines in the two patterns. This width difference arises from the different spectral composition of the primary beam in the two cases; in the X-ray case the primary beam is the CuKα line which has a spectral width of 0·005 Å and in the neutron case it is reflected from the (200) planes of a Cu single crystal with a mosaic spread of $\frac{1}{2}^\circ$; the in-pile collimation matches this approximately so that the wavelength spread in the primary beam is 0·02 Å. Thus it is clear that the resolution obtainable in neutron powder diffraction is nearly an order of magnitude less than that obtainable with X-rays. This limits the use of the method to materials with small unit cells as otherwise too much overlap between reflections occurs.

5.11 Intensity of powder diffraction

The intensity diffracted into a powder ring depends on Q, the integrated intensity per unit volume (§3.8) and on the volume of illuminated sample oriented so as to diffract. This volume depends on the multiplicity p, which

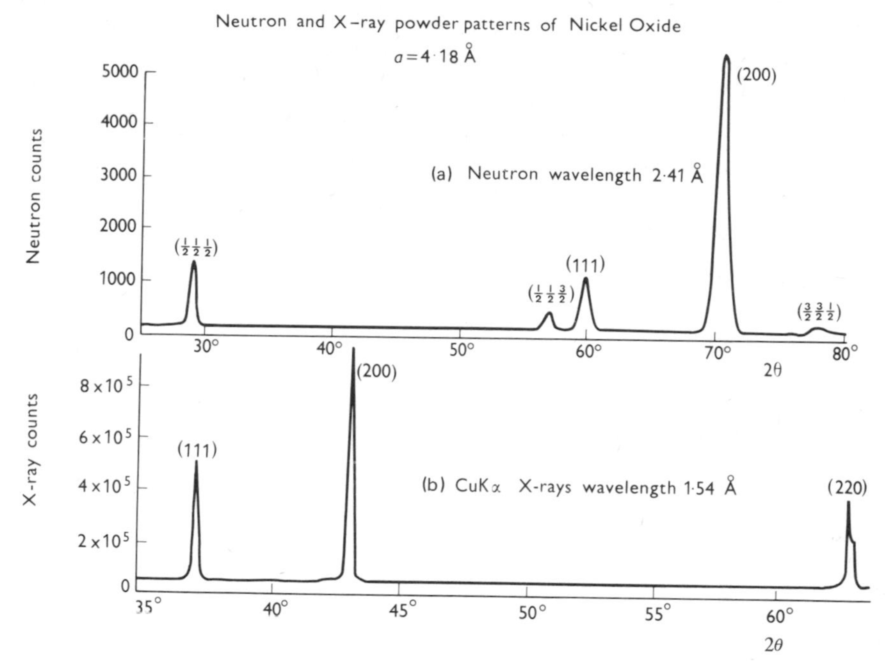

Figure 5.14 Diffractometer traces obtained from powdered NiO at room temperature, (a) using neutrons, $\lambda = 2{\cdot}41$ Å. (b) using X-rays, $\lambda = 1{\cdot}54$ Å. Note the extra peaks in the neutron diffraction pattern which arise from magnetic scattering by the antiferromagnetic array of Ni^{2+} ions.

is the number of planes in the form $\{hkl\}$ giving rise to the ring, the probability that any one of these planes will be correctly oriented ($\frac{1}{2}\cos\theta$) and the effective illuminated volume. In the case of a flat sample, thick enough to absorb the whole beam, the effective illuminated volume is $S_0/2\mu$ where S_0 is the area of the beam and μ the linear absorption coefficient. The integrated intensity *per unit length* of the diffracted ring will contain a factor $1/(\sin 2\theta)$ inversely proportional to its circumference and will therefore be proportional to

$$S_0 pQ/\mu \sin\theta \tag{5.9}$$

where Q is defined by Equ. 3.25. The intensities of the lines in a powder pattern may be measured rather easily from the data obtained from a diffractometer by integrating the area under eack peak and subtracting an integrated background measurement. The integrated background is

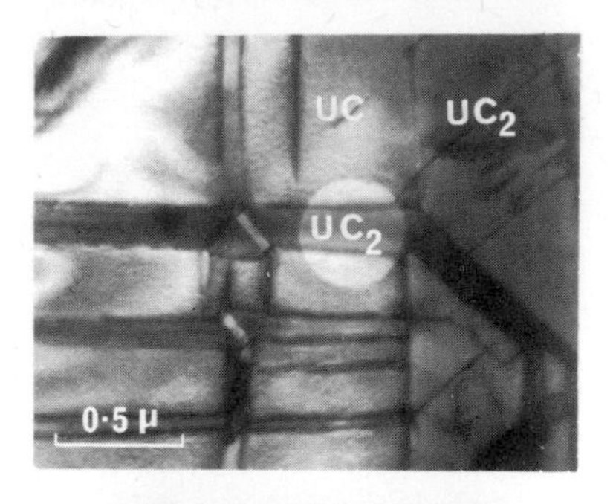

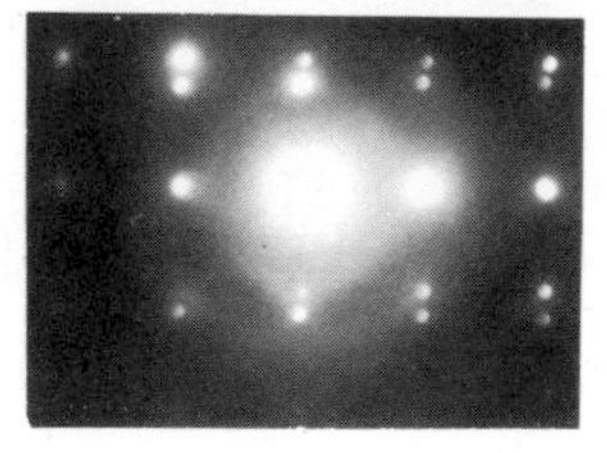

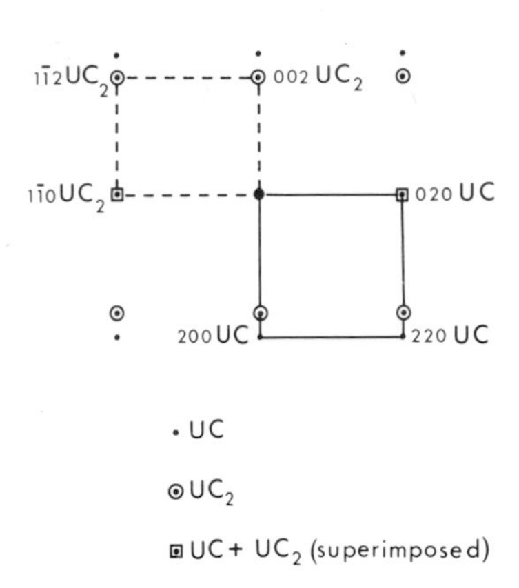

Figure 5.15 Electron microscopy and electron diffraction. The upper photograph shows an electron micrograph of a thin plate containing two uranium carbide phases; the light area is covered by the electron beam. The central picture is the corresponding diffraction pattern which contains spectra from both phases. The lower diagram shows how the reciprocal lattices of the two phases combine to produce the observed pattern which enables the relative orientations of the two single crystal phases to be established. (Photograph by kind permission of Dr B. L. Eyre, AERE Harwell. See *Phil. Mag.* (1964) 9, 545.)

estimated from the intensity on either side of the peak. The intensities of the lines on a powder photograph are not so easily found. They may be estimated visually by comparison with an intensity scale or deduced from photometric measurements of the photographic density, using an experimental calibration of density against the number of photons incident on the film.

5.12 Methods for electron diffraction

The methods used for electron-diffraction measurements are rather different from those used with either X-rays or neutrons, both because of the much higher absorption of electrons in solid specimens and because the wavelengths of the electrons used are very much shorter. It was shown in Chapter 4 that the form factor for electron scattering had much the same angular dependence as that for X-rays and had fallen to a small fraction of its maximum value at $\sin \theta/\lambda = 1{\cdot}0$. Since the wavelengths of electrons used are of the order 0·05 Å, it can be seen that almost all the electron scattering will occur at Bragg angles of less than 5°. For this reason the transmission method is the most appropriate and, because of the high absorption, the specimens must be in the form of thin flakes or films through which the electrons can pass. A simplifying feature occurs because the length of the reciprocal lattice vectors become very small in comparison with the radius of the Ewald sphere, so that near the origin of reciprocal space the sphere approximates to a plane. If this plane is perpendicular to any of the principal zones in the crystal, then the reciprocal lattice points representing planes in the zone will touch the surface of the sphere, since all have a finite extent because of the wavelength spread of the electron beam. Hence, for a single orientation of the crystal, a whole zone of reflections can be recorded. An electron-diffraction camera designed to obtain photographs using this method was illustrated in Fig. 2.4 and Fig. 5.15 shows a photograph so obtained. The reflections are very easily indexed because the photograph can be interpreted simply as a projection of the reciprocal lattice points on to the film through the centre of the Ewald sphere.

6

The Structures of the Elements

6.1 Introduction

In this chapter we shall consider the quantum mechanical interaction between like atoms and shall use this to discuss the crystal structures of the elements. Crystal structures are most easily visualised as the arrangements of hard spheres in contact. The justification for this concept lies in the strong short-range repulsive potential which exists between atoms. This enables an atomic radius to be defined which depends mainly on the range of the repulsive potential. It is however modified to a small extent by cohesive forces which depend upon the particular environment. The cohesive forces, in their turn, are determined by the electronic configuration of the element.

6.2 The periodic classification of the elements

It will be assumed that the reader is familiar with the interpretation of the Periodic Table in terms of atomic structure. The most stable state of an atom is that in which the electrons occupy the lowest permitted energy levels and this is termed the *ground state*. Characteristic excitation energies are required to cause an atom to change from the ground state and these excitation energies may be greater or less than those commonly involved in chemical reactions. In the latter case, the electronic structure of the atom in a compound may be different from the free-atom ground state listed in Table 6.1.

Many different tabular arrangements of the elements can be devised to emphasise different relationships between their properties.

The Periodic Table shown in Table 6.2 enables the elements to be classified into four types:

Type I—The inert gases which have a complete octet of electrons in the outermost s and p orbitals. This configuration is characterised by extreme stability and high ionisation potentials. Until very recently, the inert gases were believed to form no chemical compounds (but see §7.13).

Table 6.1 The electronic configurations of the ground states of the elements

Period, element and atomic number			K	L		M			N			
			1s	2s	2p	3s	3p	3d	4s	4p	4d	4f
1	H	1	1									
	He	2	2									
2	Li	3		1								
	Be	4		2								
	B	5		2	1							
	C	6	2	2	2							
	N	7		2	3							
	O	8		2	4							
	F	9		2	5							
	Ne	10		2	6							
3	Na	11				1						
	Mg	12				2						
	Al	13				2	1					
	Si	14				2	2					
	P	15	2	2	6	2	3					
	S	16				2	4					
	Cl	17				2	5					
	Ar	18				2	6					
4	K	19							1			
	Ca	20							2			
	Sc	22						1	2			
	Ti	21						2	2			
	V	23						3	2			
	Cr	24						5	1			
	Mn	25						5	2			
	Fe	26						6	2			
	Co	27						7	2			
	Ni	28	2	2	6	2	6	8	2			
	Cu	29						10	1			
	Zn	30						10	2			
	Ga	31						10	2	1		
	Ge	32						10	2	2		
	As	33						10	2	3		
	Se	34						10	2	4		
	Br	35						10	2	5		
	Kr	36						10	2	6		

Type II—Atoms which are differentiated from their neighbours in the Periodic Table by the number of electrons in the outermost shell. Such elements are characterised by a rapid change from metallic properties at the left of the Periodic Table (one electron outside the closed shell) to non-metallic properties on the right. The more metallic elements readily

Table 6.1—*continued*

Period, element and atomic number			K	L	M	N				O				P		
						4s	4p	4d	4f	5s	5p	5d	5f	6s	6p	6d
5	Rb	37								1						
	Sr	38								2						
	Y	39						1		2						
	Zr	40						2		2						
	Nb	41						4		1						
	Mo	42						5		1						
	Tc	43						5		2						
	Ru	44						7		1						
	Rh	45	2	8	18	2	6	8		1						
	Pd	46						10								
	Ag	47						10		1						
	Cd	48						10		2						
	In	49						10		2	1					
	Sn	50						10		2	2					
	Sb	51						10		2	3					
	Te	52						10		2	4					
	I	53						10		2	5					
	Xe	54						10		2	6					
6	Cs	55												1		
	Ba	56												2		
	La	57										1		2		
	Ce	58							2					2		
	Pr	59							3					2		
	Nd	60							4					2		
	Pm	61							5					2		
	Sm	62							6					2		
	Eu	63	2	8	18	2	6	10	7	2	6			2		
	Gd	64							7			1		2		
	Tb	65							9					2		
	Dy	66							10					2		
	Ho	67							11					2		
	Er	68							12					2		
	Tm	69							13					2		
	Yb	70							14					2		
	Lu	71							14			1		2		

lose electrons in their outermost shell to form positive ions, e.g. sodium forms Na^+, calcium forms Ca^{++}. The elements in Groups 6 and 7 on the other hand readily form negative ions by acquiring electrons thus completing the stable closed-shell configuration: chlorine goes to Cl^- and oxygen to O^{2-}.

Type III—Atoms forming a transition series which are differentiated from their neighbours by the number of electrons in the next to outermost

Table 6.1–*continued*

Period, element and atomic number			K	L	M	N	O				P			Q
							5s	5p	5d	5f	6s	6p	6d	7s
6	Hf	72							2		2			
	Ta	73							3		2			
	W	74							4		2			
	Re	75							5		2			
	Os	76							6		2			
	Ir	77							8		2			
	Pt	78	2	8	18	32	2	6	9		1			
	Au	79							10		1			
	Hg	80							10		2	1		
	Tl	81							10		2	2		
	Pb	82							10		2	3		
	Br	83							10		2	4		
	Po	84							10		2	5		
	At	85							10		2	6		
	Rn	86							10					
7	Fr	87												1
	Ra	88												2
	Ac	89											1	2
	Th	90								1			1	2
	Pa	91								2			1	2
	U	92								3			1	2
	Np	93								4			1	2
	Pu	94	2	8	18	32	2	6	10	5	2	6	1	2
	Am	95								7				2
	Cm	96								7			1	2
	Bk	97								8			1	2
	Cf	98								10				2
	Es	99								11				2
	Fm	100								12				2
	Md	101								13				2
	No	102								14				2

shell. All the elements are metallic and, like the Type II metals, readily form positive ions. However, since a number of excited states are usually available within a relatively small energy range of each other, the transition elements frequently exhibit several different ionisation states, e.g.

Co^{2+} $1s2s^2 2p^6 3s^2 3p^6 3d^7$ cobaltous ion

Co^{3+} $1s2s^2 2p^6 3s^2 3p^6 3d^6$ cobaltic ion

They are also able to form complex negative ions such as $(TiO_3)^{2-}$, $(VO_4)^{2-}$ and $(CrO_4)^{2-}$ which are found in the titanates, vanadates and chromates.

Table 6.2 The periodic classification of the elements

Period	Type II		Type III										Type II					Type I
	1A	2A	3A	4A	5A	6A	7A	8A			1B	2B	3B	4B	5B	6B	7B	0
1	H 1																	He 2
2	Li 3	Be 4											B 5	C 6	N 7	O 8	F 9	Ne 10
3	Na 11	Mg 12											Al 13	Si 14	P 15	S 16	Cl 17	Ar 18
4	K 19	Ca 20	Sc 21	Ti 22	V 23	Cr 24	Mn 25	Fe 26	Co 27	Ni 28	Cu 29	Zn 30	Ga 31	Ge 32	As 33	Se 34	Br 35	Kr 36
5	Rb 37	Sr 38	Y 39	Zr 40	Nb 41	Mo 42	Tc 43	Ru 44	Rh 45	Pd 46	Ag 47	Cd 48	In 49	Sn 50	Sb 51	Te 52	I 53	Xe 54
6	Cs 55	Ba 56	La 57	Hf 72	Ta 73	W 74	Re 75	Os 76	Ir 77	Pt 78	Au 79	Hg 80	Tl 81	Pb 82	Bi 83	Po 84	At 85	Rn 86
7	Fr 87	Ra 88	Ac 89															

Type IV																
Lanthanides	6	Ce 58	Pr 59	Nd 60	Pm 61	Sm 62	Eu 63	Gd 64	Tb 65	Dy 66	Ho 67	Er 68	Tm 69	Yb 70	Lu 71
Actinides	7	Th 90	Pa 91	U 92	Np 93	Pu 94	Am 95	Cm 96	Bk 97	Cf 98	Es 99	Fm 100	Md 101	No 102	

Type IV—A further transition series in which elements are differentiated by the number of electrons in the second to outermost shell. The lanthanide series of rare-earth elements is characterised by changes in the 4f shell of electrons ($4f^1$–$4f^{14}$) and the actinide series by similar changes in the occupation of 5f levels. The highly shielded nature of these shells ensures that the chemical similarity between members is great. This similarity is not found in their magnetic properties, since these are related primarily to the number of unfilled 4f and 5f states respectively.

6.3 The hydrogen molecule ion, H_2^+

The simplest case we can consider in which more than one atom is involved is the hydrogen molecule ion H_2^+, consisting of two hydrogen nuclei sharing a single electron. We shall not attempt to obtain an exact solution to the Schrödinger equation in this case but will use the variational technique to illustrate the approximate solution of the problem.

The Schrödinger equation

$$\mathcal{H}\psi = E\psi \tag{6.1}$$

is satisfied by eigenfunctions ψ of the Hamiltonian $\mathcal{H}$ with energies E. If we multiply both sides of the equation by ψ^* and integrate over the whole of space we obtain

$$E = \int \psi^* \mathcal{H} \psi \, \mathrm{d}\tau \bigg/ \int \psi^* \psi \, \mathrm{d}\tau$$

and if ψ happens to be the ground state then E is the minimum possible energy. Any approximate wavefunction ψ_1 must therefore be such that

$$\int \psi_1^* \mathcal{H} \psi_1 \, \mathrm{d}\tau \bigg/ \int \psi_1^* \psi_1 \, \mathrm{d}\tau = E_1 \geqslant E$$

and the nearer E_1 is to E the closer is our approximate function ψ_1 to the true ground-state wavefunction.

In choosing an approximate wavefunction to describe the electron in the hydrogen molecule ion we are guided by the symmetry of the problem. The most obvious characteristic of a wavefunction in a diatomic molecule is that it should be bi-centric; however, in the region of either nucleus the significant parts of the Hamiltonian are exactly those terms which comprise the Hamiltonian of an electron isolated on that nucleus. Thus we expect the wavefunction to have characteristics of both ψ_A and ψ_B, the individual atomic wavefunctions for nucleus A and nucleus B. It therefore seems reasonable to write the approximate molecular wavefunction as a linear combination of the appropriate atomic wavefunctions. This is known as the linear combination of atomic orbitals or L.C.A.O. approximation. Thus we write ψ_1 as

$$\psi_1 = \psi_A + \lambda\psi_B$$

where ψ_A is an atomic wavefunction centred on atom A and ψ_B an equivalent wavefunction centred on atom B. Then

$$E_1 = \frac{\int(\psi_A^* \mathcal{H}\psi_A + 2\lambda\psi_A^* \mathcal{H}\psi_B + \lambda^2\psi_B^* \mathcal{H}\psi_B)\,d\tau}{\int(\psi_A^*\psi_A + 2\lambda\psi_A^*\psi_B + \lambda^2\psi_B^*\psi_B)\,d\tau}$$

$$= (E_A + 2\lambda\beta + \lambda^2 E_B)/(1 + 2\lambda S + \lambda^2) \qquad (6.2)$$

where β has been written for the *resonance integral* $\int \psi_A^* \mathcal{H}\psi_B\,d\tau$ and S for the *overlap integral* $\int \psi_A^* \psi_B\,d\tau$. The minimum value of E_1 is obtained by choosing λ so that $\partial E_1/\partial\lambda = 0$; since $E_A = E_B$ we obtain $\lambda^2 = 1$ or $\lambda = \pm 1$. The energies corresponding to these two solutions are

$$E\pm = \frac{E_A \pm \beta}{1 \pm S} \qquad (6.3)$$

and the wavefunctions are $\psi\pm = \psi_A \pm \psi_B$. To a first approximation the overlap integral S may be put equal to zero so that

$$E\pm = E_A \pm \beta \qquad (6.4)$$

(β is normally negative and therefore the ψ_+ state has lower energy). Equations 6.3 and 6.4 show how the two degenerate atomic wavefunctions ψ_A and ψ_B combine to give two non-degenerate wavefunctions ψ_+ and ψ_- whose energies differ by an amount which depends on the magnitude of the resonance integral β. Figure 6.1a shows the energies of the molecular orbitals ψ_+ and ψ_- for H_2^+ as a function of the nuclear separation. The coulomb energy e^2/R has been included to give the total energy of the molecule. The variation of the charge density $\psi\psi^*$ along the molecular axis is illustrated in Fig. 6.1b for both the bonding (ψ_+) and antibonding (ψ_-) states.

6.4 The application of molecular orbital theory to the elements

We have seen in the preceding section that when we bring two similar atoms together atomic orbitals centred on one atom combine with atomic orbitals centred on the other to give pairs of orbitals ψ_+ and ψ_-; one of these, the bonding orbital ψ_+, has a lower energy than the original atomic orbitals whereas the antibonding ψ_- orbital has a higher energy. The dif-

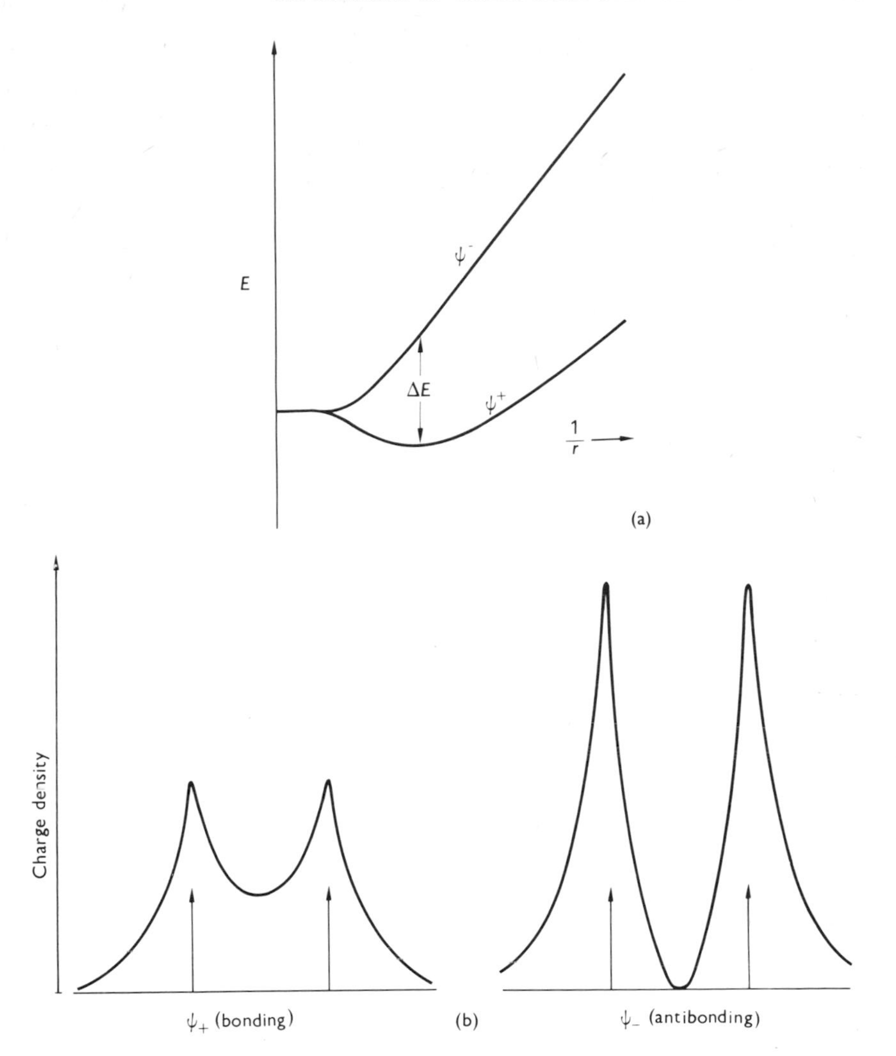

Figure 6.1 (a) The energies (E) of the molecular orbitals ψ^+ and ψ^- for H_2^+ as a function of nuclear separation r. (b) the variation of the charge density $\psi\psi^*$ along the molecular axis for the bonding (ψ_+) and antibonding (ψ_-) states; the vertical arrows indicate the positions of the nuclei

ference in energy between the bonding and antibonding states is determined by the appropriate resonance and overlap integrals and at large separation the energies of both states tend to that of the parent atomic orbitals. At small separation both energies rise steeply because of internuclear repulsion. The rise is particularly dramatic for the antibonding orbitals because of the increase in the overlap integral as the atoms approach one another. This effect is responsible for the instability of molecules such as He_2 in which the antibonding orbitals would have to be occupied because the original atoms contain fully occupied atomic orbitals.

This repulsive interaction between closed shells of atomic orbitals is also responsible for the steeply rising repulsive potential between many-electron atoms and underlies the success of the hard-sphere model of crystal structures. The onset of the steeply rising potential enables characteristic radii to be assigned to the elements; these are plotted against atomic number in Fig. 6.2.

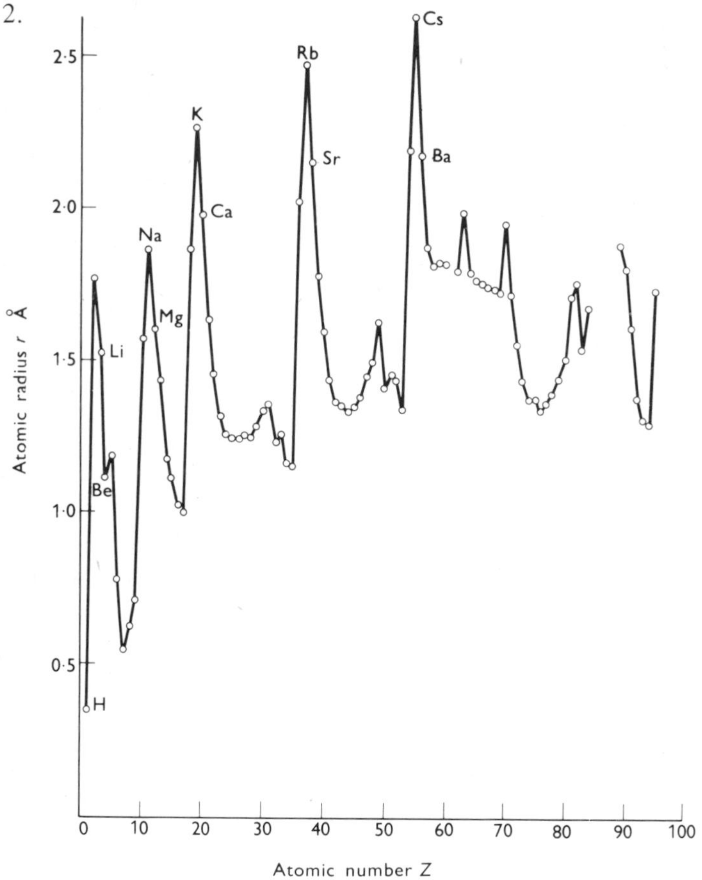

Figure 6.2 The characteristic radii of the elements.

It can be seen from this figure that the radii associated with elements having similar electronic structures increase with increasing atomic number, for example the series Be, Mg, Ca, Sr, and Ba. It should be noted however that the change in radius between the later members of the series is approximately constant and is not related to the number of intervening elements.

6.5 The electron wave functions in a solid*

In the preceding section we have seen how it is possible to obtain approximate wavefunctions for a system consisting of two atoms. Clearly the problem of calculating the wavefunction in a solid is much more complicated, not least because the potential in which the electrons move is itself determined by the wavefunction. An exact solution is not possible, but considerable physical insight into the structurally important features of wavefunctions in solids may be obtained from an extension of the L.C.A.O. approximation. We saw in Section 6.3 that the effect of bringing two similar atoms together was to produce two diatomic orbitals $\psi_A \pm \psi_B$ separated in energy from the original atomic orbital ψ_A or ψ_B. If we continue the process, building up a solid by adding further atoms, each new atom adds one more orbital with energy in the range between E_+ and E_- and slightly modifies those of the previous set. In the limit this process results in a band of allowed energies corresponding to each atomic orbital. This band contains one state per atom for each of the original atomic orbitals. The process of formation of such a band is illustrated in Fig. 6.3a, and the dependence of the band widths on internuclear distance in Fig. 6.3b.

If the degree of overlap between orbitals of adjacent atoms is small so that the widths of the bands are less than the energy separation of the original atomic orbitals then the bands corresponding to the orbitals will remain separated. In all cases electrons in the closed shells of atoms belong to this non-overlapping group and give rise to narrow 'atomic' bands with energies and wavefunctions little perturbed from those in the isolated atoms. If the converse conditions exist, the bands may overlap in energy and the resulting molecular orbitals are combinations of both types of atomic orbital. It is a well-known result of quantum mechanics that when states of a system of similar energies and suitable symmetry exist the true eigenstates of the system are linear combinations of these states, thus an s

* A detailed discussion of the material of this and the following section will be found in the companion text in this series, *The Electronic Structures of Solids*.

and a p state of similar energy combine to form two hybrid states $a\psi_s + b\psi_p$ and $a\psi_s - b\psi_p$. The wavefunctions of these two states are illustrated in Fig. 6.4.

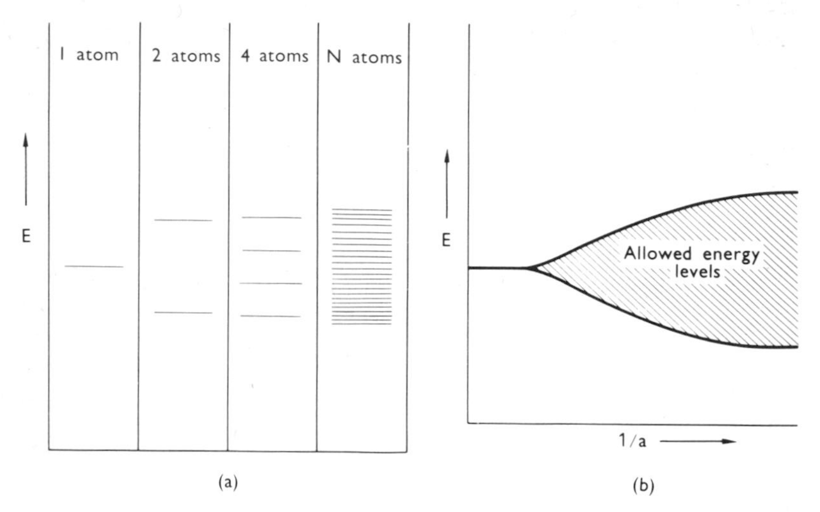

Figure 6.3(a) The process of band formation by the addition of further atoms to build up a solid from a single atom. (b) The dependence of the band widths on internuclear distance.

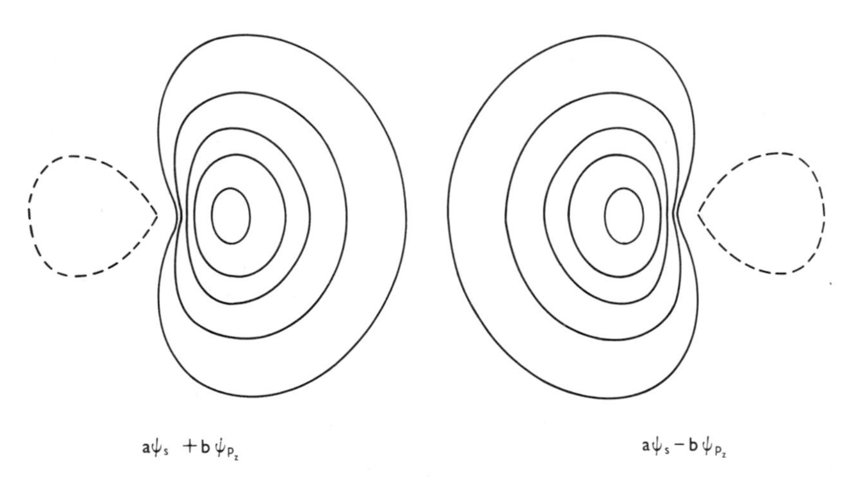

Figure 6.4 The nybrid states $a\psi_s + b\psi_p$ and $a\psi_s - b\psi_p$. Contours are at equal arbitrary intervals, negative levels being shown dashed.

It can be seen that the wavefunctions are highly directional in character and this has important consequences for the stability of structures. The effect of hybridisation on the band structure of a solid can be illustrated by the case of silicon in which the 3s and 3p bands have similar energies. The effect of decreasing internuclear distance is illustrated in Fig. 6.5. The top of the 3s band and the bottom of the 3p band gradually approach one another in energy but instead of crossing at the point A interaction occurs between the s and p states. The interaction results in the formation of sp^3 hybrid bonding and antibonding bands separated by an energy gap.

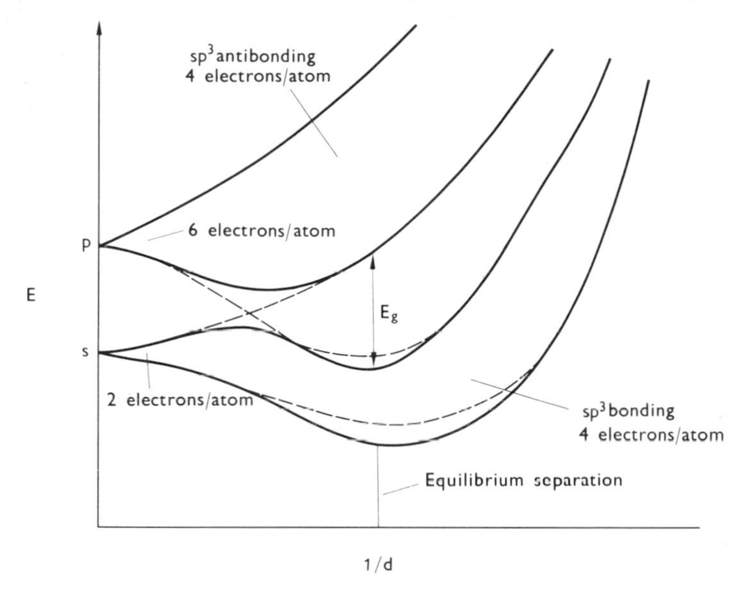

Figure 6.5 The effect of decreasing the internuclear distance on the energy bands of silicon. The full lines give the limiting energies of the hybrid bands and the dashed lines the approximate form of the unhybridised s and p bands. The unhybridised p band in silicon would only be partly filled with 4 electrons per atom. The resultant sp^3 bonding band is completely filled and is separated by the energy gap E_g from the unfilled sp^3 antibonding band.

6.6 The effect of the lattice periodicity on the wavefunctions

In Section 6.5 we obtained a qualitative picture of the band structure of a solid by considering it to be a molecule consisting of a very large number of atoms. This approach took no account of the periodicity of the crystal lattice. An alternative way of attempting to find solutions to the wave equation in a solid exploits the periodic nature of the lattice and hence of the electronic potential. Bloch (1928) proved that in a periodic potential the wavefunctions have the form

$$\psi_k = U_k(\boldsymbol{r}) \exp(i\boldsymbol{k} \cdot \boldsymbol{r}) \tag{6.5}$$

where $U_k(\boldsymbol{r})$ is a function having the lattice periodicity.

The Bloch function depicted in Equ. 6.5 can be labelled by the vectors $\boldsymbol{k}$ and these can be conveniently represented as vectors in k-*space*—a space obtained directly from the reciprocal lattice (§3.4) by multiplying all dimensions by 2π. The influence of the lattice periodicity on the Bloch functions may be demonstrated by considering the function $\psi_k(\boldsymbol{l} + \boldsymbol{r})$ where $\boldsymbol{k} = \boldsymbol{g} + \boldsymbol{k}'$; $\boldsymbol{g}$ is a lattice vector of $\boldsymbol{k}$ space and $\boldsymbol{l}$ is a lattice repeat in the crystal.

$$\begin{aligned}\psi_k(\boldsymbol{l}+\boldsymbol{r}) &= U_k(\boldsymbol{l}+\boldsymbol{r}) \exp(i(\boldsymbol{g}+\boldsymbol{k}') \cdot (\boldsymbol{l}+\boldsymbol{r})) \\ &= U_k(\boldsymbol{r}) \exp(i(\boldsymbol{g}+\boldsymbol{k}') \cdot \boldsymbol{r}) \exp(i(\boldsymbol{g}+\boldsymbol{k}') \cdot \boldsymbol{l}) \\ &= \psi_k(\boldsymbol{r}) \exp(i\boldsymbol{k}' \cdot \boldsymbol{l})\end{aligned}$$

since $\boldsymbol{g} \cdot \boldsymbol{l} = 2\pi n$ with n being an integer. Thus the function ψ_k behaves like a Bloch function with wavevector $\boldsymbol{k}' = \boldsymbol{k} - \boldsymbol{g}$. The wavevector of a Bloch function is therefore not unique; corresponding to each function there are an infinite number of equivalent $\boldsymbol{k}$ vectors differing from it by lattice vectors $\boldsymbol{g}$ of $\boldsymbol{k}$ space. For most purposes it is convenient to consider a set of non-equivalent $\boldsymbol{k}$ vectors. Such a set may clearly be mapped within a single unit cell of $\boldsymbol{k}$ space since no two points within the cell are separated by a lattice vector and every point outside can be related to one inside by translation by such a vector. The Brillouin zones are alternative volumes of $\boldsymbol{k}$ space which contain the set of non-equivalent $\boldsymbol{k}$ vectors once only. The first Brillouin zone is the smallest closed polyhedron enclosing the origin of $\boldsymbol{k}$ space, formed by planes which perpendicularly bisect the lattice vectors $\boldsymbol{g}$. Higher Brillouin zones are formed by subsequent closed polyhedra. Thus the first Brillouin zone contains those parts of $\boldsymbol{k}$ space which are nearer to the origin than to any other lattice point. Since all lattice points are equivalent the zone contains all non-equivalent $\boldsymbol{k}$ vectors and hence its volume is the same as that of the unit cell of $\boldsymbol{k}$ space. The geometrical forms of the first Brillouin zones of the face-centred and body-centred cubic lattices are illustrated in Fig. 6.6.

The total number of non-equivalent $\boldsymbol{k}$ vectors is determined by the boundary conditions at the surface of the crystal. For a crystal containing $N_1 \times N_2 \times N_3$ unit cells the values of $\boldsymbol{k}$ which satisfy simple periodic boundary conditions are given by

$$N_1\boldsymbol{a}\,.\,\boldsymbol{k} = 2\pi n_1; \quad N_2\boldsymbol{b}\,.\,\boldsymbol{k} = 2\pi n_2; \quad N_3\boldsymbol{c}\,.\,\boldsymbol{k} = 2\pi n_3$$

where n_1, n_2, n_3 are integers. In terms of the lattice vectors of $\boldsymbol{k}$ space, $2\pi\boldsymbol{a}^*$, $2\pi\boldsymbol{b}^*$, $2\pi\boldsymbol{c}^*$, these conditions can be written

$$\boldsymbol{k} = \frac{n_1}{N_1}(2\pi\boldsymbol{a}^*) + \frac{n_2}{N_2}(2\pi\boldsymbol{b}^*) + \frac{n_3}{N_3}(2\pi\boldsymbol{c}^*)$$

thus the allowed values of $\boldsymbol{k}$ are radius vectors of a sub-lattice whose basis vectors are $2\pi\boldsymbol{a}^*/N_1$, $2\pi\boldsymbol{b}^*/N_2$, $2\pi\boldsymbol{c}^*/N_3$. The volume of $\boldsymbol{k}$ space per $\boldsymbol{k}$ vector is $1/N_1N_2N_3$ times the volume of the unit cell of $\boldsymbol{k}$ space, and hence the number of non-equivalent $\boldsymbol{k}$ vectors is equal to the number of primitive unit cells in the crystal.

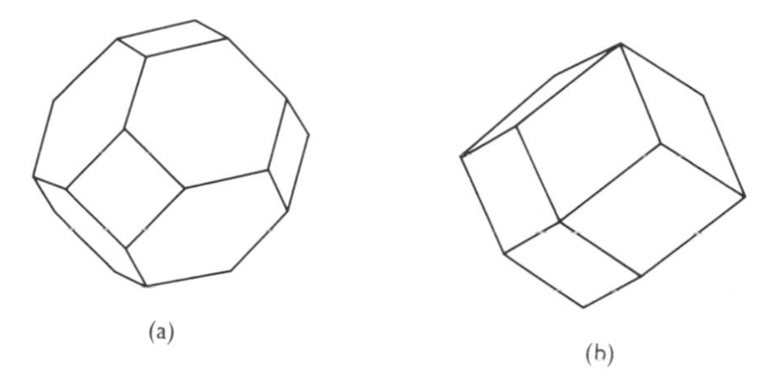

Figure 6.6 The first Brillouin zones in (a) the f.c.c. and (b) the b.c.c. lattices.

In the free electron case, for which the periodic potential is zero, the Bloch functions ψ_k are plane waves $\psi_k = e^{i\boldsymbol{k}\,.\,\boldsymbol{r}}$. The electrons have kinetic energy only and hence their energies E_k are given by $E_k = h^2k^2/2m$ so that there is a parabolic dependence of E on $\boldsymbol{k}$ (Fig. 6.7a).

The effect of a periodic potential is to introduce discontinuities in the energy at values of $\boldsymbol{k}$ corresponding to Brillouin zone boundaries (Fig. 6.7b). Electrons with $\boldsymbol{k}$ values just inside the zone boundaries have their energies reduced and those with $\boldsymbol{k}$ just outside have their energies enhanced, relative to the parabolic dependence. These changes in energy arise because electrons with wavevectors near the critical values interact strongly with the crystal lattice. Electrons whose wavevectors correspond to points on the zone boundaries can undergo Bragg reflection by the lattice, and this would result in transfer of energy between the electron and the lattice due to momentum conservation. Thus electrons with these wavevectors can have no continuing existence in the lattice. The dependence of E on $\boldsymbol{k}$ in a periodic potential may also be illustrated for successive bands within the first Brillouin zone. The result is termed the reduced zone scheme, and is illustrated in Fig. 6.7c.

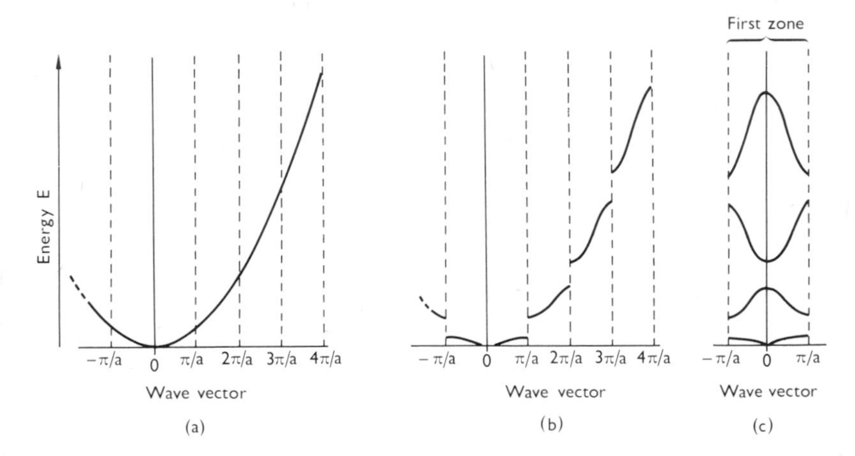

Figure 6.7 Energy (E) versus momentum (k) curves. (a) Parabolic, corresponding to a free electron. (b) Modified by discontinuities introduced by the crystal lattice. (c) The reduced zone scheme corresponding to (b).

The number of states available to electrons within a given energy range, $N(E)\,\mathrm{d}E$, may be plotted against the energy E to give the density-of-states curve. The forms of these curves are strongly dependent on the strength of the periodic potential, which dictates the magnitude of the energy discontinuities. They also depend on the geometric form of the Brillouin zones, which may determine whether there is a completely forbidden energy range when all directions in $\boldsymbol{k}$ space are considered. Figure 6.8 is a schematic density of states curve for a substance in which the first two zones overlap in energy.

6.7 The inert gases: close-packed structures

The elements He, Ne, Ar, Kr, Xe and Rn are gases which can only be liquefied and finally solidified at very low temperatures. These elements have closed outer shells of s and p electrons whose wavefunctions do not overlap significantly even in the solid state. The resulting fully occupied

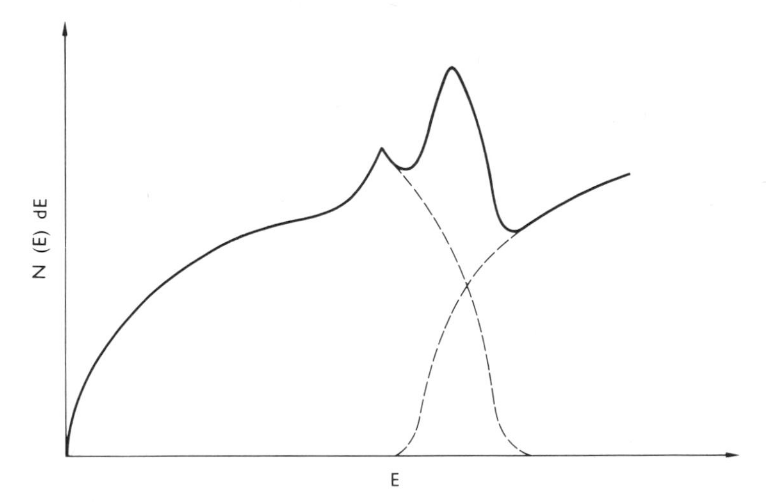

Figure 6.8 A schematic density of states curve, N (E) dE, plotted as a function of the energy (E) for the case of overlapping zones.

bands are extremely narrow compared to the energy difference between bands. In effect, the band structure consists of a highly degenerate set of discrete orbital levels.

Cohesion in the condensed state results from weak attractive forces, a satisfactory explanation of which was first given by London (1930). London showed that the cohesive force arises from the dynamic polarisation of an atom which results from its zero-point motion. The inert gas atoms do not possess permanent dipole moments but they may have an instantaneous moment since at any instant the electron distribution need not have the full spherical symmetry of its time-average. The instantaneous dipole moment of one atom will polarise its neighbour in just the manner described by Debye (1920) for atoms and molecules exhibiting a permanent dipole moment. Since the interaction is dipolar, the interatomic potential will depend on r^{-6} and is therefore short-ranged. This and other weak attractive interactions are loosely termed Van der Waal's forces commemorating Van der Waal's introduction (1873) of such forces into an equation of state for real gases.

The Van der Waal's bond is non-directional and the inert gases crystallise so that the atoms are equally spaced. Figure 6.9 illustrates the derivation of the two most symmetrical arrangements of close-packed spheres of equal radii. In both cases the packing fraction is $4\left(\frac{4\pi r^3}{3}\right)\Big/\left(\frac{4r}{\sqrt{2}}\right)^3$, i.e. 0·741.

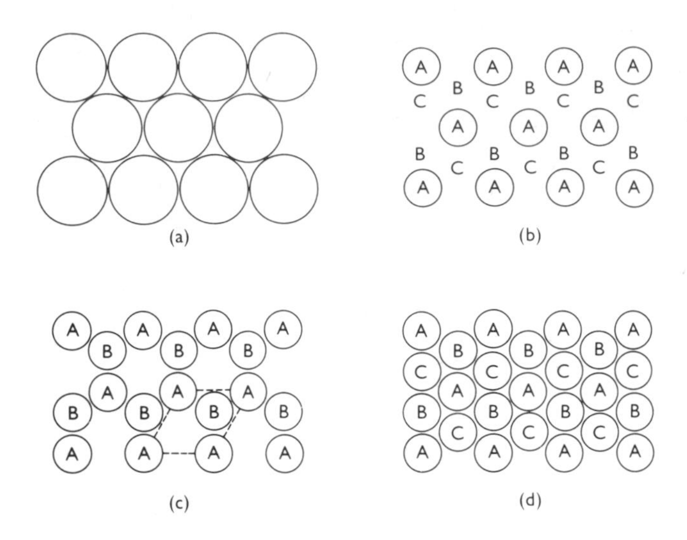

Figure 6.9 The close-packing of spheres of equal radius. (a) A single layer. (b) The positions B and C of two sets of hollows in the surface of the single layer. (c) Hexagonal close packing made up of successive A and B layers. The unit cell is indicated by dashed lines. (d) Cubic close packing in which the layers repeat ABC ABC and lie perpendicular to ⟨111⟩.

Helium solidifies under high pressure in the hexagonal close-packed structure (h.c.p.); the remaining inert gases are cubic close-packed (c.c.p.) in the solid state. In both structures each atom is surrounded by twelve nearest neighbours at the same distance. The number of atoms which may be considered to be in 'contact' with any particular atom is said to be the *co-ordination number* (C.N.) for that atom. We may also refer to the co-ordination of an atom in terms of the polyhedron whose vertices are the centres of the co-ordinating atoms. The polyhedra corresponding to c.c.p. and h.c.p. are illustrated in Fig. 6.10. It may be noted in connection with structures assembled from close-packed layers that the stacking sequence is determined by interactions between alternate rather than adjacent layers. Such interactions are often weak so that mistakes in the sequence (termed *stacking faults*) may readily occur.

The low strength of the Van der Waal's bond is also reflected in the low heats of sublimation ranging from 0·6 kcal/mol for Ne to 2·8 for the more easily polarisable Kr (cf. 100 kcal/mol for Fe).

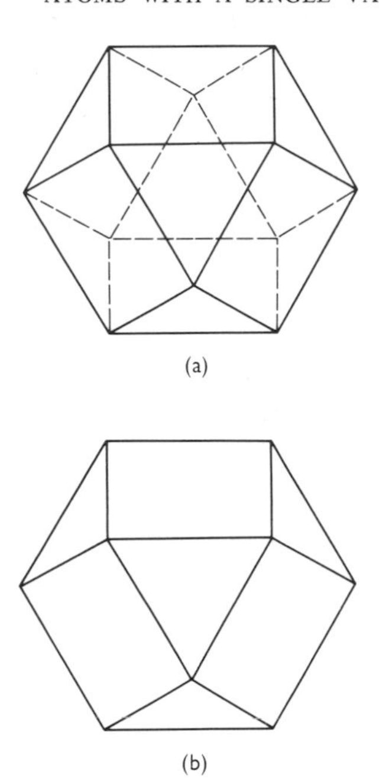

Figure 6.10 The coordination polyhedra for (a) cubic close packing; a cube octahedron where the three points lying below the median plane are rotated relative to those lying above (b) for hexagonal close packing, where the two sets of points are superimposed.

6.8 Atoms with a single valence electron: the half-filled band

We have shown (§6.6) that there are two electron states per atom available in each Brillouin zone. Atoms of such metals as the alkali metals Na, K, Cs and Rb and the noble metals Cu, Ag and Au have one electron outside a stable shell. This electron is loosely bound and the corresponding zone is half-filled. A consequence of this is that these elements are good electrical and thermal conductors at all temperatures down to absolute zero, since states are always available just above the Fermi level into which electrons can be excited. The Fermi surface is very nearly spherical and the electron states isotropic; hence the cohesive forces have little directional character. One might expect therefore close-packed structures and these are in fact found for the noble metals which are cubic (cf. Fig. 6.9). Surprisingly, the alkali metals have a body-centred cubic structure with one atom per lattice point (Fig. 6.11a). The packing fraction is lower (0·68) than for the close-packed structures (0·74). Each atom has only eight nearest neighbours but there are a further six neighbours across the faces of the cube at only 15% greater distance. Thus the b.c.c. structure is often said to have C.N. = 14 and its co-ordinating polyhedron is illustrated in Fig. 6.11b. The three structures b.c.c., h.c.p. and c.c.p. are typical of the metallic state. It may be noted that lithium and sodium have h.c.p. structures at low temperatures (Barrett, 1956).

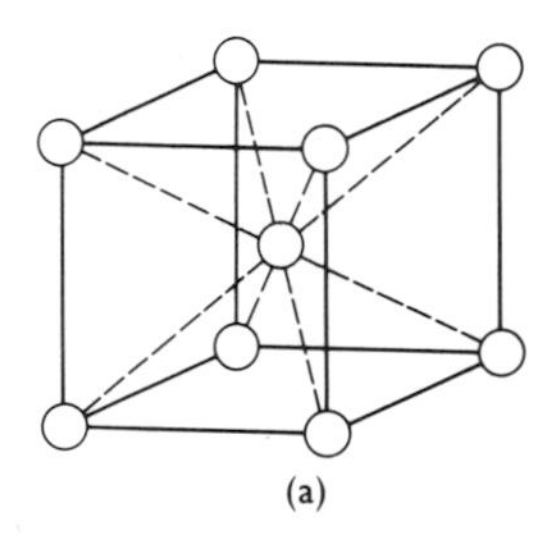

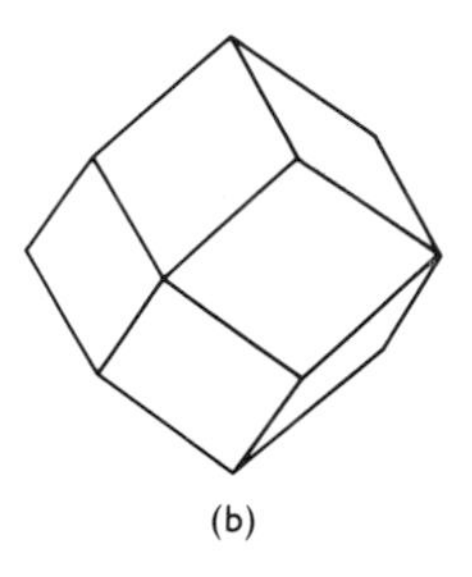

Figure 6.11 (a) The body centred cubic structure. (b) Its 14-fold coordination polyhedron which includes the second nearest neighbours; the polyhedron is a rhombic dodecahedron.

Some understanding of the difference in behaviour between the alkali and noble metals may be had by contrasting potassium with copper. In both elements the single valence electron is of 4s type, but in potassium the 3d shell is empty whereas in copper it is full. The consequent lower mean electron density in the outer regions of the alkali metal atom contributes less to the repulsive potential as the atoms approach one another and leads to the high compressibility of potassium relative to copper. As might be expected, the ionic radius of K^+ is much less than the atomic radius in the metal, but a much smaller difference exists between the radii of Cu^+ and Cu. This distinction between the alkali and noble metals led Hume-Rothery and Raynor (1938) to introduce the concept of open (compressible) and full (incompressible) metals.

6.9 The transition, lanthanide and actinide elements

The free-atom ground states of the transition elements have an incomplete d shell†, and are therefore characterised by having more than one unfilled shell. Some of the salient features of the band structure of transition metals may be illustrated by nickel. The 3d wavefunctions are more

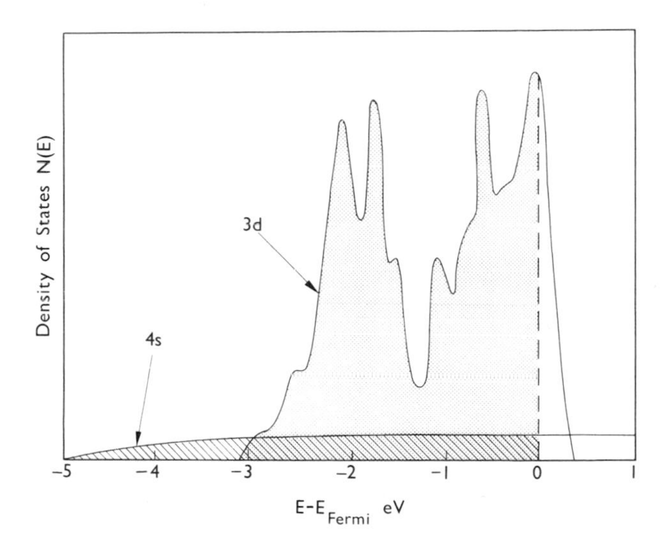

Figure 6.12 Density of states in nickel for the bands close to the Fermi energy, as calculated by Koster (1955). The Fermi level is shown as a vertical broken line.

localised than the 4s and their energies will be modified to a lesser extent in the solid. The corresponding density-of-states curve in the two bands is illustrated in Fig. 6.12, which shows that the density of states is much greater in the d band.

†In the case of palladium, the incomplete shell exists for excited states of small energy.

Figure 6.12 is a simplification in that exchange interaction has been neglected. The effect of this interaction is seen principally on the narrow 3d band, which is divided into two parts corresponding to each of the two electron spin states. The spin-up and spin-down bands overlap to some extent, and the magnetic moment is a measure of the difference between the numbers of electrons in each. The wavefunctions corresponding to

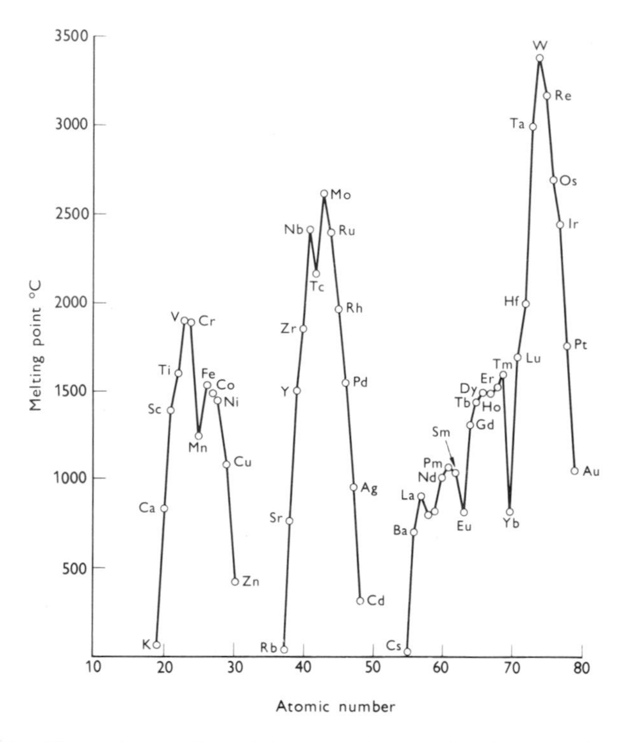

Figure 6.13 The melting points of the transition metal and rare earth elements.

states in the lower half of the d band have bonding character and contribute electron density to regions between atoms in the solid. The filling of these 'bonding' states is to be associated with the decrease in atomic radius (Fig. 6.2) and increase in stability found between the initial and final members of each transition metal series. Figure 6.13 shows how the increase in stability is reflected by the change in melting point across the three transition metal series. States in the upper half of the d band are more 'atomic' in character and may give rise to localised magnetic moments.

Table 6.3 shows the structure exhibited by the transition metals. With the exception of manganese, they are the three typical metallic structures. It is significant that there is a tendency amongst the middle members of the series to form body-centred cubic structures and this may be interpreted as a preference for high co-ordination when the d 'bonding' orbitals are nearly filled. Manganese is an extreme example of this tendency. The α-form, stable at room temperature, has a complex cubic structure containing 58 atoms/unit cell of four crystallographically non-equivalent types

Table 6.3 The structures of the transition metals

Period	Group							
	3A	4A	5A	6A	7A	8		
4	Sc A_1† A_3†	Ti A_3 A_2†	V A_2	Cr A_2 A_3	Mn complex A_1 A_2	Fe A_2 A_1	Co A_3 A_1	Ni A_1 A_3
	Y A_2	Zr A_3 A_2	Nb A_2	Mo A_2 A_3	Tc A_3	Ru A_3	Rh A_1	Pd A_1
6	La A_1 A_3	Hf A_3 A_2	Ta A_2	W A_2	Re A_3	Os A_1	Ir A_1	Pt A_1
7	Ac A_1							

†A_1, cubic close-packing; A_2, cubic body-centred; A_3, hexagonal close-packing.

with C.N. values of 16, 16, 13 and 12. The same structure occurs in alloys between two different transition metals when it is called the χ-phase (cf. §8.5).

The other groups of elements with more than one incomplete shell are the lanthanide and actinide series listed in Table 6.4. The lanthanide or rare earth series is characterised by the close-packed structures having relatively low stacking-fault energy. The cubic structure is preferred by the early members and the hexagonal structure by the later ones. The intermediate elements, Nd and Sm, have more complex structures derived from close-packed layers stacked in the order ABAC ABAC (cf. §6.7).

Few of the actinide elements possess simple metallic structures. β-Uranium is tetragonal with 30 atoms/unit cell of five unequivalent types: this structure is also found, like that of α-manganese, amongst alloys between two different transition metals when it is known as the σ-phase (cf. §8.5). α-Plutonium is monoclinic with eight non-equivalent sets of atoms. Many other complex structures exist in the series and some of these are also found amongst alloys.

Table 6.4 The structures of the lanthanide and actinide elements

Ce	Pr	Nd	Pm	Sm	Eu	Gd	Tb	Dy	Ho	Er	Tm	Yb	Lu
A_1† A_3	A_1 C†	C	–	C	A_2	A_3†	A_3	A_3	A_3	A_3	A_3	A_1	A_3
Th	**Pa**	**U**	**Np**	**Pu**	**Am**	**Cm**	**Bk**	**Cf**	**Es**	**Fm**	**Md**	**No**	
A_1 A_2†	C	C A_2	C	C A_1 A_2									

†A_1, cubic close-packing; A_2, cubic body-centred; A_3, hexagonal close-packing; C, complex structure or structures.

6.10 Polyelectronic metals in Group II

The elements we have considered in the previous two sections have atomic orbitals which are incompletely filled and it is clear why they have metallic properties. The alkaline earth metals (Group IIA), on the other hand, contain two valence electrons per atom which are just sufficient to fill the 2s band. Their conductivity must result from an overlap between 2s and 2p bands in the solid state. The wavefunctions will however still be essentially s-like (non-directional) in character and this is borne out by the fact that these elements adopt the structure characteristic of the true metals.

The elements zinc, cadmium and mercury (Group IIB) are also metallic in character, though the hexagonal structures of zinc and cadmium are distorted from h.c.p. and have c/a ratios of between 1·8 and 1·9 compared to the 1·633 appropriate to true close-packing. As a consequence of this distortion, each atom has six neighbours at a shorter distance than the remaining six which occur in the layers above and below it. This preference for C.N. 6 is more pronounced in mercury, which has atoms at the corners of a rhombohedral unit cell. The lack of complete isotropy in the structures of these elements shows us that the wavefunctions involved in the bonding cannot be entirely s-like.

6.11 Directional bonding

The elements so far considered have structures which may be understood largely in terms of a spherically symmetric potential associated with wave-functions having little directional character. The remaining elements which occupy Groups IIIB–VIIB in the Periodic Table (Table 6.5) have structures which cannot be understood in this simple way.

Table 6.5 The B sub-group elements of Groups II to VII

Period	Group					
	2B	3B	4B	5B	6B	7B
2		B	C	N	O	F
3		Al	Si	P	S	Cl
4	Zn	Ga	Ge	As	Se	Br
5	Cd	In	Sn	Sb	Te	I
6	Hg	Tl	Pb	Bi	Po	At

From Group IIIB onwards, valence electrons partially occupy the p levels with the important result that the wavefunction no longer possesses spherical symmetry around each atom site. The aspherical character may be further increased by hybridisation when bands overlap (§6.5). In these conditions the energy of a particular arrangement may be critically dependent on the relative dispositions of neighbouring atoms and the resulting interatomic forces are non-central. The effect is more marked for elements of low atomic number. The valence electrons of atoms of high atomic number have energies in a range in which there are many different electronic states available, and in the combined states which result most of the directional character is averaged out. For the same reason, it is unlikely that a band gap will exist at the Fermi energy so that most heavy elements exhibit metallic conduction.

The bonding and antibonding orbitals associated with s-state electrons have already been considered in the discussion of the hydrogen molecule ion (§6.3). The six atomic p orbitals in a diatomic system separate into two with axes along the line joining the two atoms and four with axes perpendicular to this direction. The molecular orbitals are similarly of two types; σ, formed by combining the orbitals parallel to the axis of the molecule, and π, formed from the others. Thus there can be seen to be eight orbitals available which are the bonding and antibonding s and σ, and two degenerate π combinations.

A possible energy level scheme for these orbitals is shown in Fig. 6.14a. This scheme is only valid when the original s and p levels are well separated in energy. If this is not the case hybridisation may occur (see §6.5) and Fig. 6.14b shows a possible energy level scheme when there is strong sp hydridisation. The sp hybrid orbitals are illustrated on Fig. 6.4. Other types of hybridisation can occur; sp^3 (cf. §6.5) and sp^2 which has three lobes making angles of 120° with each other. The energy separation of

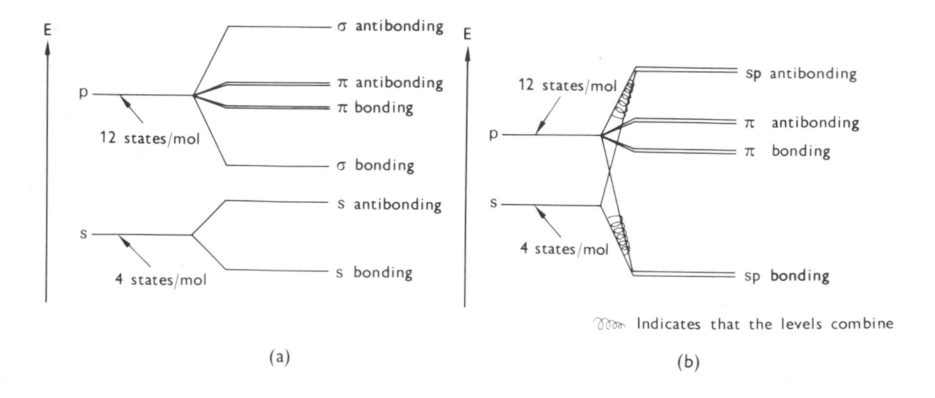

Figure 6.14 Possible energy level schemes for electrons in a diatomic molecule. (a) When the s and p orbitals are well separated in energy. (b) When the s and p levels are close together in energy and sp hybrids are formed.

bonding and antibonding combinations of a given set of orbitals depends on the degree of overlap, being largest for those which overlap most. Hence the appearance of the σ_a (antibonding) and sp_a orbitals at the highest energies on Fig. 6.14a and b. This enables us to understand the tendency of all atoms, save those of the inert gases, to form diatomic molecules, since when the antibonding states are not completely filled the net energy of the molecule is less than that of the two separate atoms.

It remains to be considered whether in any particular case the diatomic molecule is energetically more or less favourable than some other bonded configuration. It can be seen from Figs. 6.14a and b that no energy premium arises if both antibonding and bonding combinations of a given set of orbitals are occupied. In the case of s and p orbitals only, if there are more than four electrons per atom the final configuration can be deduced from the energy of the *unoccupied antibonding orbitals*, whereas with less than four electrons per atom it is the *filled bonding orbitals* that should be considered.

For Group VII atoms (seven valence electrons per atom) only one of the antibonding states is unoccupied. The maximum overlap in this orbital can be achieved along the axis of a diatomic molecule and these are therefore stable (F, Cl, Br and I). The situation is more complicated when more than one orbital must be considered, since overlap may then occur with orbitals belonging to more than one neighbour. This happens in S, Se, and Te where the atoms link into chains, but in oxygen the diatomic molecule is still energetically favourable. The tendency outlined above forms the basis of what has been known as the $(8 - N)$ rule: this suggests that the structure of an element will be stable if the number of closely bonded neighbours is equal to $(8 - N)$, where N is the number of its group. Evidence for this assertion is provided mainly by the structures of the elements in Groups IVB to VIIB, though the deviation from the ideal c/a ratio in the h.c.p. structures of Zn and Cd (§6.10) has been attributed to this effect.

6.12 Group IIIB elements

The first member of this group is boron. It has at least two polymorphs, each of which is built of icosahedral frameworks of atoms stacked in a different way. The simplest form, which is rhombohedral with 12 atoms per unit cell, forms transparent red crystals. Its structure is illustrated in Fig. 6.15. The same icosahedral framework of atoms is found in other boron-rich compounds such as UB_{12}.

The β-rhombohedral form of boron is a semiconductor with a band gap of 1.4 eV. It is known that boron forms three-centre bonds and it is generally thought that these are responsible for the lack of metallic behaviour.

In contrast to boron, aluminium behaves as a true metal. It has the c.c.p. structure and de Haas van Alphen measurements show that the Fermi surface is nearly spherical.

The remaining elements in the group, gallium, indium and thallium are also metallic conductors but like boron the structures of Ga and In reveal directional bonding. The heaviest member, thallium, has b.c.c. and h.c.p. modifications.

6.13 Group IVB elements

The first few elements in the group (Table 6.5) can all crystallise in the diamond structure illustrated in Fig. 6.16a.

It is the stable modification of Si and Ge and it is also adopted by tin at low temperatures (grey tin) and by carbon under pressure (diamond). The strength and stability of the structure is due to the formation of sp^3 bonding and antibonding bands (cf. §6.5), the electron density of which is directed along the tetrahedral directions to neighbouring atoms. The number of electrons (four per atom) is just sufficient to fill the bonding (valence) band and the antibonding (conduction) band is unoccupied. The energy gap between the two bands is sufficient to ensure that the materials are insulators at low temperatures. Silicon and germanium are intrinsic semiconductors however, since their energy gap is of order 1 eV.

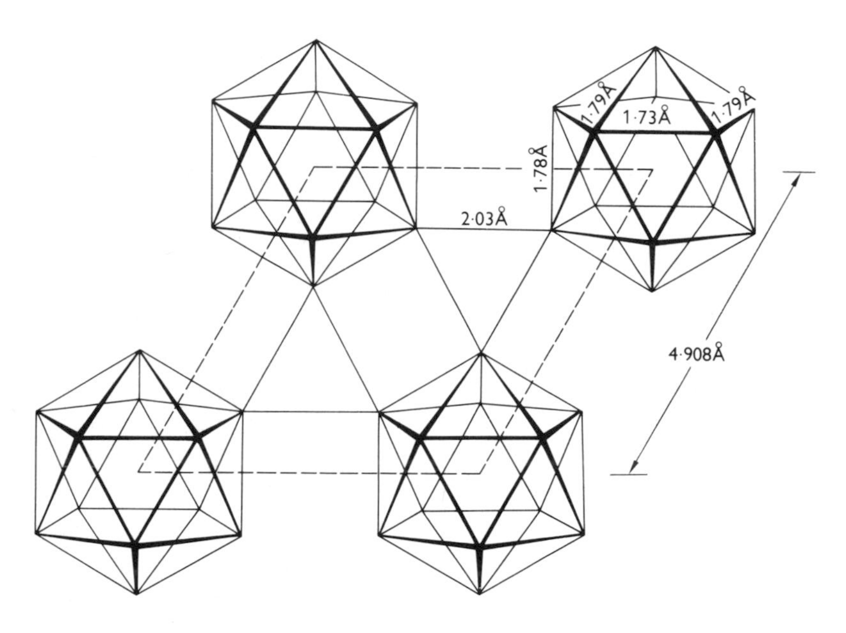

Figure 6.15 The structure of rhombohedral boron projected onto the (001) plane of the related hexagonal cell with a 4·91 Å and c = 12·57 Å. The bond distance to boron atoms in icosehedra above and below in the c direction is 1·71 Å.

Carbon is dimorphous and also occurs as graphite, whose structure is illustrated in Fig. 6.16b. The hexagonal nets of carbon atoms extend in infinite sheets and have a C—C distance of 1·42 Å (cf. 1·54 Å in diamond). The layers of carbon atoms are superimposed with a much longer C—C distance of 3·35 Å. This remarkable anisotropy of the bonding results from the formation of the sp^2 hybrid orbitals, which form σ bonds to the other carbon atoms within a single sheet. The remaining electrons (one per carbon) occupy π orbitals extending above and below the plane of the sheet.

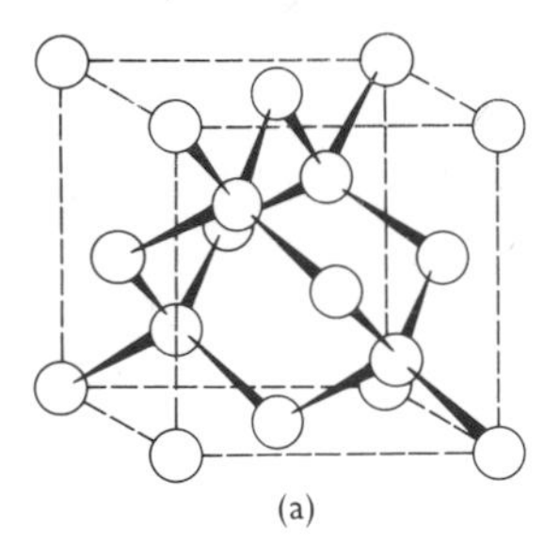

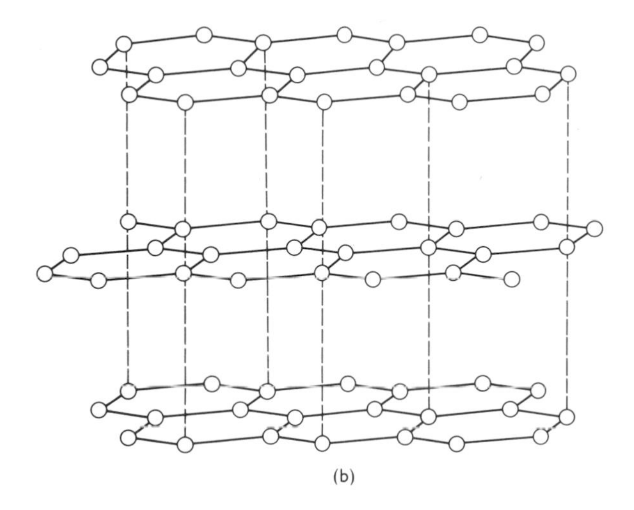

Figure 6.16 (a) The diamond structure (cubic, a = 3·56 Å). (b) The graphite structure (hexagonal, a = 2·46 Å, c = 6·80 Å). Broken lines link atoms of successive sheets which are in the same vertical columns.

The non-localisation of these electrons is evident from the relatively high electrical conductivity of single-crystal graphite in directions parallel to the layers compared to the perpendicular direction. The weakness of the interlayer bonding gives rise to the extremely pronounced cleavage of the material and hence to its lubricating properties. The strength of the intralayer bonds is utilised in composites containing carbon fibres. Each fibre is obtained by the graphitisation of a synthetic yarn and the axis of the fibre lies in the hexagonal planes of the structure. White tin has a unique structure showing directional character but is a metallic conductor. The heaviest member of the series, lead, is like thallium in being a metal with, in this case, the c.c.p. structure. The band structure of lead is not simple however, and its Fermi surface has a complicated topology.

6.14 Group VB elements

Nitrogen forms a diatomic molecule which exists in the solid, liquid and gas phases (§6.11). The liquefaction and solidification of nitrogen result from Van der Waal's forces. The crystal structure, which is common to other bimolecular solids, is a hexagonal close-packing of the dumb-bell shaped molecules.

Phosphorus also forms molecules, P_4, of tetrahedral form which are found in the vapour, in solution and in the yellow crystalline modification.

The black or 'metallic' form of phosphorus together with the heavier members of the group all have structures formed from puckered layers in which each atom has three close neighbours. In the structure common to As, Sb and Bi the layers are hexagonal and are stacked together to form a rhombohedral lattice (Fig. 6.17).

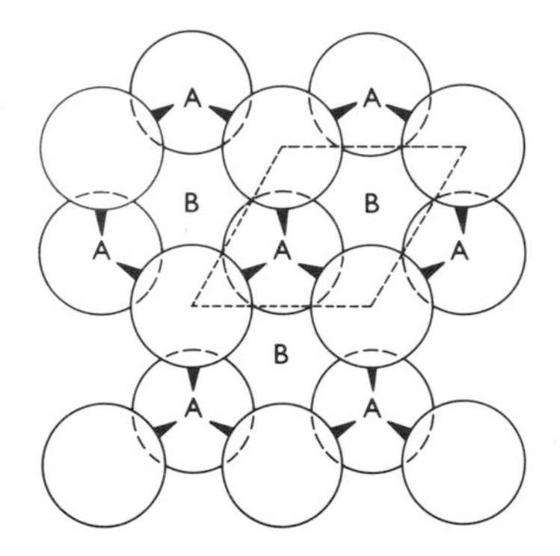

Figure 6.17 A plan view of a puckered sheet of atoms as found in the structures of arsenic, antimony and bismuth. The lower atoms (A) in the next layer fall vertically above the points B in the layer below. The basal plane of the hexagonal unit cell is dashed.

The electrical resistivity of As, Sb and Bi decreases with decreasing temperature which suggests metallic conduction; however in arsenic at 4·2 K the resistivity perpendicular to the layers is an order of magnitude greater than that parallel to the layers. This behaviour is similar to that of graphite (§6.13) and suggests that the 'metallic' conduction arises from delocalisation of the π-bonding electrons.

6.15 The chalcogenides (elements in Group VIB)

Structurally, oxygen is similar to nitrogen (§6.14) but there are three known modifications of the crystal structure in at least one of which, the α-phase, the unquenched p-orbital moment leads to antiferromagnetism.

The remaining elements in the group form bonds to two neighbours giving rise to a number of structures containing rings (sulphur and selenium) or chains (a polymorph of selenium and tellurium). The directional character of the orbitals involved leads to an inter-bond angle of about 105° and consequently the rings are buckled and the chains twisted into spirals. The structure of selenium which contains such spirals is illustrated in Fig. 6.18a.

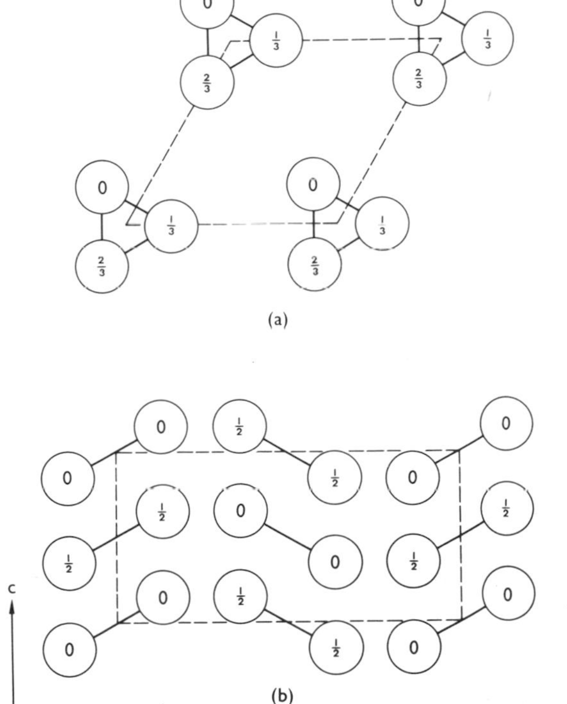

Figure 6.18 (a) Projection on (001) of the structure of hexagonal selenium (a = 4·37 Å, c = 4·96 Å). (b) Projection on (100) of the structure of iodine (orthohombic a = 7·27 Å, b = 9·79 Å, c = 4·79 Å). The I-I distance in the molecule is 2·70 Å whereas the shortest distance between molecules is 3·54 Å.

6.16 The halogens

These all form diatomic molecules. The solidification temperature increases with increasing atomic number, showing a corresponding increase in the strength of the Van der Waal's bond. Iodine, which is a solid at room temperature, is black and lustrous indicative of a small band gap; it becomes metallic under pressure. Its structure is illustrated in Fig. 6.18b.

7

Polar Structures

7.1 Introduction

A new feature which is introduced when studying structures containing more than a single type of atom is the essential asymmetry of the inter-atomic interaction. We may illustrate this by consideration of the series of binary compounds formed by pairs of elements:

Si
Al . . . . P
Mg . . . . S
Na Cl

The crystal structure of silicon and its energy bands have been described in Section 6.13. The four electrons per atom completely fill the sp^3 bonding band and there is a gap of 0·8 eV between the top of this band and the bottom of the conduction band. AlP, with the same number of electrons per atom (e/a), forms a related structure (of the zinc blende type) in which the atomic sites are the same as in silicon but the four neighbours of any one atom are of the other sort (see Fig. 7.1). The essential difference in the electronic configuration arises from the difference in nuclear charges. The electron potential relative to that in silicon, is raised near to the aluminium atoms and lowered near to the phosphorus atoms which leads to a shift in charge density towards the phosphorus atom. This effect is seen yet more strongly as we progress to MgS and NaCl. Whilst e/a is still 4, the asymmetry of the charge distribution reaches a point in NaCl where the ions Na^+ and Cl^- may be distinguished. Since these ions possess the inert gas configur-ation the electrons will occupy a series of narrow, filled 'ionic' bands.

When ions are formed, the cohesion of the structure arises mainly from the electrostatic interaction between dissimilar ions. Both MgS and NaCl have the same structure, which is illustrated in Fig. 7.2a. In it, each ion is octahedrally co-ordinated by six ions of the opposite polarity and the structure is exhibited by a large number of ionic compounds.

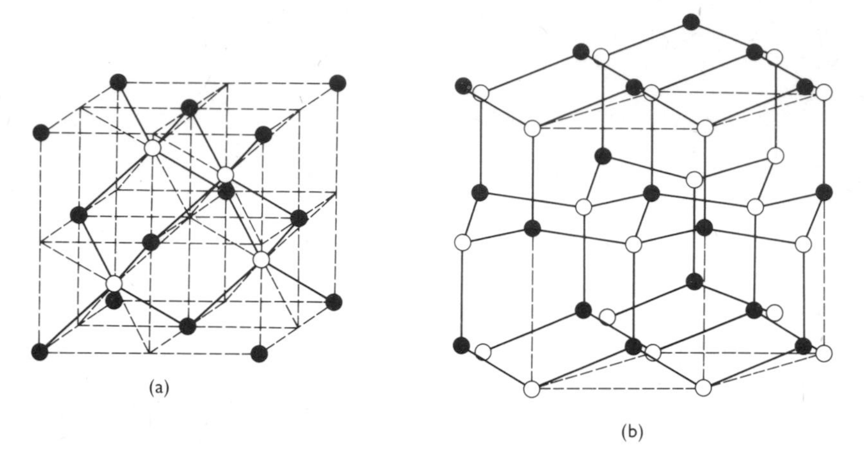

Figure 7.1 (a) The structure of zinc blende (cubic a = 5·42 Å) and (b) wurtzite (hexagonal a = 3·84 Å, c = 5·18 Å).

7.2 The effect of ionic size on the stability of structures

All the alkali halides except CsCl, CsBr and CsI possess the sodium chloride structure. These three salts, however, have the even simpler structure in which ions of one type occupy the corners of the cubic unit cell and ions of the other type occur at the body-centre position. This arrangement results in a C.N. value of 8 and also occurs in many binary alloys when its stability does not derive from ion formation (cf. §8.9). The zinc blende structure described in the previous section is particularly associated with compounds between elements in Group N and those in Group $(8 - N)$, e.g. GaP, InAs, BeTe and CuCl, as is the alternative structure formed by ZnS in its wurtzite modification. This latter structure, illustrated in Fig. 7.1b, has the same tetrahedral co-ordination as zinc blende but the tetrahedra are stacked together in a different way so as to have hexagonal symmetry. Among other compounds with this structure are AlN, BeO, ZnO and AgI.

Consideration of the cell dimensions of the alkali halides reveals a systematic expansion with increasing atomic number of either the cation or the anion. A set of characteristic radii for the ions may then be deduced on the assumption that the cell dimension is determined by the contact between oppositely charged ions each with a more or less constant size.

The values for the alkali halides and halogen ions are given in Table 7.1.

Table 7.1 Ionic radii (Å) for ions with C.N. = 6

Li^+	0·68	F^-	1·36
Na^+	0·97	Cl^-	1·81
K^+	1·33	Br^-	1·95
Rb^+	1·47	I^-	2·16
Cs^+	1·67		

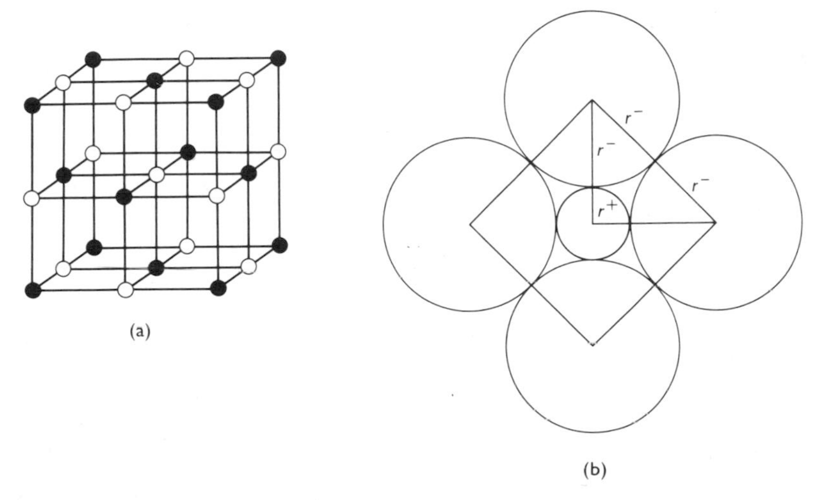

Figure 7.2 (a) The sodium chloride structure and (b) a plan of the structure in the (001) plane which enables the limiting radius ratio to be calculated.

If we accept the concept of a characteristic ionic size there will exist a *radius ratio* r^+/r^- between cations and anions below which it is impossible to fit six anions around a cation whilst maintaining anion–cation 'contact'. The critical radius ratio then corresponds to the situation illustrated in Fig. 7.2b, i.e.

$$r^+ + r^- = \sqrt{2}\, r^-$$

$$r^+/r^- = (\sqrt{2} - 1) = 0{\cdot}414$$

Similar considerations applied to C.N. = 4 and C.N. = 8 enable us to compile Table 7.2.

The effective radius exhibited by an ion depends on the co-ordination in which it occurs. Lowering the co-ordination results in an apparent contraction of the ionic radius, since the strength of the electrostatic interactions increases the number of oppositely charged neighbours decreases. Ionic radii are usually quoted for C.N. = 6; the approximate corrections for different ionic co-ordinations are C.N. = 8, +3% and C.N. = 4, –5%.

Table 7.2 Radius ratio ranges for the stability of simple ionic structures

Co-ordination	Typical structure	r^+/r^- for stability
8	CsCl	$>0{\cdot}732$
6	NaCl	$0{\cdot}414 < r^+/r^- < 0{\cdot}732$
4	ZnS	$0{\cdot}225 < r^+/r^- < 0{\cdot}414$

7.3 Ionic polarisation

The process of ion formation described in Section 7.1 can be partially reversed in compounds where the size difference between the anions and the cations is large. The high relative field strength at the 'surface' of the small cation tends partly to reverse the process of charge separation and re-introduces a directional character to the bond. In such cases *polarisation* is said to occur: a small cation is an efficient polariser and a large anion is highly polarisable. The polarisability of an ion can be deduced from measurements of refractive index in the optical region (e.g. Tessman *et al.*, 1953). Experimental values for the halides and alkali metal ions are given in Table 7.3 and illustrate the principles outlined above. The presence of polarisation is reflected in the 'ionic' band structure, the band associated with the polarisable electrons being significantly broadened.

Table 7.3 The electronic polarisabilities of alkali halide ions in units of $cm^3 \times 10^{-24}$ (Tessman *et al.*, 1953).

F^-	0·65	Li^+	0·03
Cl^-	2·97	Na^+	0·41
Br^-	4·17	K^+	1·33
I^-	6·44	Rb^+	1·98
		Cs^+	3·34

Evidence for increasing directional character in the bonding is provided by the occurrence of salts crystallising with the zinc blende or wurtzite structures when the radius ratio of the ions would indicate a preference for the NaCl structure. Examples are CuF, CuCl, CuBr and AgI.

7.4 Ionic compounds with composition AX_2

Before passing on to consider more complex ionic compounds between more than two elements it is worth describing some of the more important structures exhibited by binary compounds occurring at the composition AX_2. The co-ordination of the anion and cation cannot be reciprocal and the larger negative ions usually determine how close the packing can be. We therefore expect that the highest co-ordination will be 8, 4 corresponding to $r^+/r^- > 0{\cdot}732$ as in CsCl.

This co-ordination is found in CaF_2, the structure of which is illustrated in Fig. 7.3a. Figure 7.3b shows the structure of TiO_2 in its rutile modification in which each titanium atom is co-ordinated by six oxygen atoms and each oxygen atom by three titanium atoms. Finally, Fig. 7.3c shows the cristobalite modification of quartz, SiO_2, in which the silicon atom has a C.N. value of 4 and the oxygen atom a C.N. value of 2.

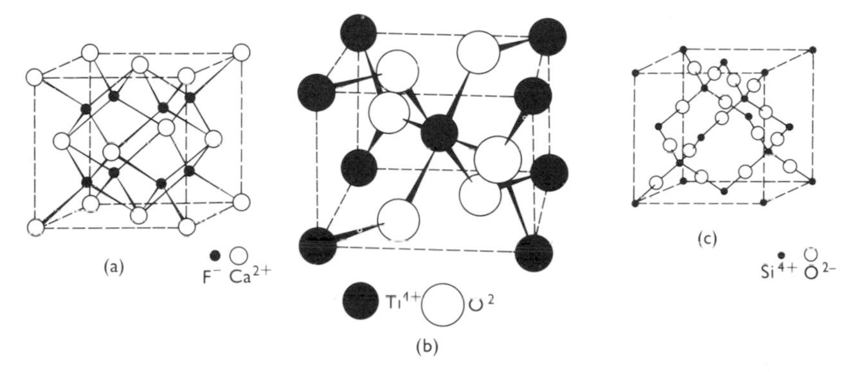

Figure 7.3 The structure of (a) calcium fluoride (cubic a = 5·45 Å), (b) rutile, TiO_2 (tetragonal a 4·58 Å, c = 2·95 Å) and (c) the crystobalite modification of quartz (cubic a = 7·12 Å).

In all these structures the bonding is relatively homogeneous, but this is not the case in the layer structures typified by the cadmium halides. In both the rhombohedral structure of CdI_2 and the hexagonal one of $CdCl_2$ layers of cadmium atoms are octahedrally co-ordinated by halide atoms to form a sandwich (Fig. 7.4).

The two cells result from stacking the hexagonal layers in the sequence AB AB (hexagonal) or ABC ABC (rhombohedral). The energy difference between these two arrangements is small and PbI_2 occurs in both forms. The cohesion between successive sandwich layers is weak, being residual in character, so that the structure exhibits pronounced cleavage parallel to

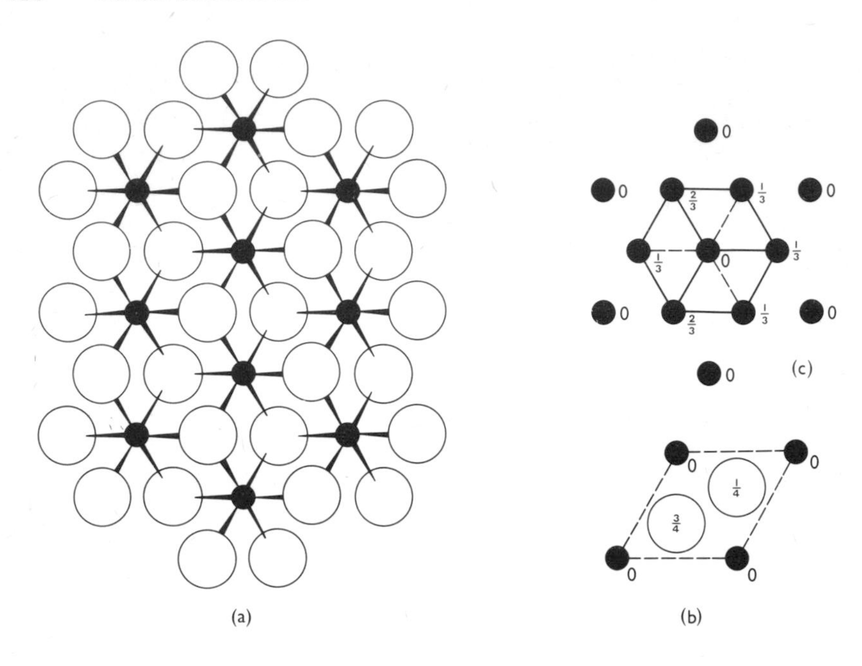

Figure 7.4 The structure of cadmium chloride and cadmium iodide. (a) Shows the basic layer common to both structures. The solid circles represent cadmium atoms octahedrally coordinated by six anions, three above and three below. A direct superposition of layers forms the hexagonal unit cell (a = 4·24 Å, c = 6·84 Å) of cadmium iodide shown in (b). Displacement of the cadmium atoms in successive layers as in (c) results in the rhombohedral unit cell of cadmium atoms are shown together with their fractional heights along the [111] trigonal direction which corresponds with the c axis of a hexgonal cell with a = 3·99 Å, c = 17·7 Å.

the layers. The bonding within each layer is strong with a large covalent contribution and the structure occurs between elements where the X ion is readily polarisable, e.g. Cl. Br, I, S, and the cation is strongly polarising, e.g. Ag, Zn, Cd, Co, Ni, Fe.

7.5 Polymorphism in ionic compounds

The equilibrium configuration of a polyphase material at the temperature T and pressure P results in a minimum in the Gibbs free energy G, where

$$G = U + PV - TS$$

U is the internal energy, V is the volume and S is the entropy of the system.

Two phases may exist in equilibrium if they have the same value of G per unit mass. The internal energy of a crystal structure is a function of all the postional parameters of the atoms, and it may exhibit a number of minima as illustrated schematically in Fig. 7.5. Each configuration will have its own particular value of S and V per unit mass, so that the stable form depends on the particular conditions of temperature and pressure.

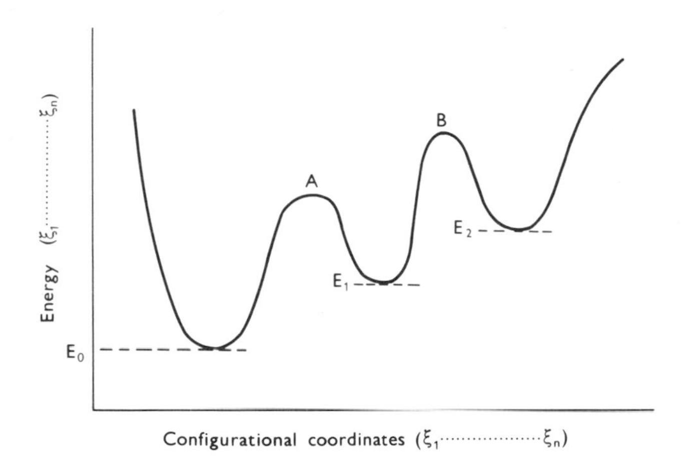

Figure 7.5 A hypothetical variation of the internal energy of a crystal as a function of the atomic positional parameters.

These considerations apply to systems in thermodynamic equilibrium which may not in practice be achieved. The transformation from one phase to another may involve the formation of a substantially new structure or it may only require a small distortion of the original structure. The probability of the former process occurring is negligible except at high temperature, whereas the second process may be relatively probable at all temperatures. It is therefore likely that the transition may be arrested by quenching in the former case, but this may not be possible in the latter, *displacive*, type of transition. The achievement of even a displacive transition involves intermediate states of higher energy and if this energy difference, *the activation energy*, exceeds that associated with thermal vibrations, the transition is unlikely to occur.

Many ionic substances occur naturally as minerals which have been formed under widely different conditions of temperature and pressure. Such materials frequently exhibit polymorphism as the non-equilibrium configurations have been retained by quenching. The polymorphs of SiO_2 and TiO_2 illustrate some features common to many polymorphic systems. In both compounds the same basic co-ordination polyhedron of anions about the cation is found in all structures, but the arrangement of the polyhedra differs between the polymorphs. There is also a tendency toward higher symmetry as the formation temperature of the polymorph increases. The modifications of quartz provide an illustration: the Si–O–Si bond is bent in the low-temperature modifications α- and β-quartz, but the mean position for the oxygen atom is midway between the silicon atoms in tridymite and cristobalite. The latter structure is illustrated in Fig. 7.3c; the structure of tridymite is related to wurtzite (Fig. 7.1b) in the same way as cristobalite is related to zinc blende. Each Zn and S atom is replaced by silicon and the oxygen atoms are situated midway between each pair of silicon atoms.

The trend towards higher symmetry with higher temperature of stability is also evident in TiO_2 since the high-temperature modification, rutile, has more symmetrical (TiO_6) octahedra than occur in either brookite or anatase. The tetragonal structure of rutile is illustrated in Fig. 7.3b.

7.6 Mixed oxides: the perovskite and spinel structures

As a first example of an important class of compounds which require three or more different types of atom for their formation, we will examine two structures commonly found amongst the oxides, sulphides, selenides and tellurides having two different cation species.

The simpler of these structures is illustrated in Fig. 7.6, which shows the unit cell of the perovskite structure containing one formula unit ABO_3.

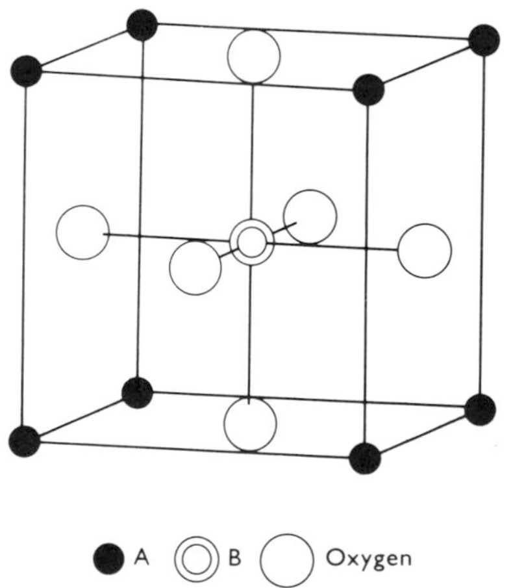

Figure 7.6 The unit cell of the perovskite structure.

Cation A has a C.N. value of 12 whereas cation B has a C.N. value of 6; the A atom is therefore the larger and the A and oxygen atoms together constitute a cubic close-packed arrangement. Examples of this structure are given in Table 7.4. There are many structures related to the simple perovskite arrangement. The basic co-ordination numbers remain the same, but the polyhedra are distorted, e.g. $BaTiO_3$, $CaWO_3$ etc. In $BaTiO_3$ successive

Table 7.4 Some compounds with either the ideal or a slightly distorted perovskite structure

$NaNbO_3$	$CaTiO_3$	$CaSnO_3$	$BaPrO_3$	$YAlO_3$	$KMgF_3$
$KNbO_3$	$SrTiO_3$	$SrSnO_3$	$SrHfO_3$	$LaAlO_3$	$PbMgF_3$
$NaWO_3$	$BaTiO_3$	$BaSnO_3$	$BaHfO_3$	$LaCrO_3$	$KNiF_3$
	$CdTiO_3$	$CaCeO_3$	$BaThO_3$	$LaMnO_3$	$KZnF_3$
	$PbTiO_3$	$SrCeO_3$		$LaFeO_3$	
	$CaZrO_3$	$BaCeO_3$			
	$SrZrO_3$	$CdCeO_3$			
	$BaZrO_3$	$PbCeO_3$			
	$PbZrO_3$				

Some compounds with the spinel structure.

$BeLi_2F_4$	$MgCr_2O_4$†	$MgFe_2O_4$‡	$FeNi_2O_4$	$MgGa_2O_4$‡
$MoNa_2F_4$	$MnCr_2O_4$	$TiFe_2O_4$‡	$GeNi_2O_4$	$ZnGa_2O_4$
WNa_2O_4	$FeCr_2O_4$	$MnFe_2O_4$	$FeNi_2S_4$	$CaGa_2O_4$
$ZnK_2(CN)_4$	$CoCr_2O_4$	$FeFe_2O_4$‡	$NiNi_2S_4$	$MgIn_2O_4$‡
$CdK_2(CN)_4$	$NiCr_2O_4$†	Fe_2O_3	$MgRh_2O_4$†	$CaIn_2O_4$
$HgK_2(CN)_4$	$ZnCr_2O_4$†	$CoFe_2O_4$	$ZnRh_2O_4$†	$MnIn_2O_4$
$TiMg_2O_4$‡	$CdCr_2O_4$†	$NiFe_2O_4$	$TiZn_2O_4$‡	$FeIn_2O_4$‡
VMg_2O_4†	$MnCr_2S_4$	$CuFe_2O_4$‡	$SnZn_2O_4$‡	$CoIn_2O_4$‡
$SnMg_2O_4$	$FeCr_2S_4$	$ZnFe_2O_4$†	$MgAl_2O_4$	$NiIn_2O_4$‡
MgV_2O_4†	$CoCr_2S_4$	$CdFe_2O_4$†	$SrAl_2O_4$	$CdIn_2O_4$
FeV_2O_4	$CdCr_2S_4$	$AlFe_2O_4$	$CrAl_2O_4$	$HgIn_2O_4$
ZnV_2O_4†	$HgCr_2S_4$†	$PbFe_2O_4$	$MoAl_2O_4$	$MnGa_2O_4$
	$ZnCr_2Se_4$	$MgCo_2O_4$	$FeAl_2O_4$	
	$CdCr_2Se_4$†	$TiCo_2O_4$	$CoAl_2O_4$	
	$TiMn_2O_4$	$CoCo_2O_4$	$NiAl_2O_4$†	
	$MgMn_2O_4$	$CuCo_2O_4$	$CuAl_2O_4$	
	$CuMn_2O_4$	$ZnCo_2O_4$	$ZnAl_2O_4$†	
	$CuCr_2O_4$	$SnCo_2O_4$†	Al_2O_3	
	$CuCr_2S_4$	$CoCo_2S_4$	$ZnAl_2S_4$	
	$CuCr_2Se_4$	$CuCo_2S_4$	$CuRh_2S_4$	
	$CuCr_2Te_4$	$CaFe_2O_4$		
		$Ag_{1/2}In_{1/2}Cr_2S_4$		
		$Cu_{1/2}Fe_{1/2}Cr_2S_4$		
		$Cu_{1/2}In_{1/2}Cr_2S_4$		
		$Cu_{1/2}In_{1/2}Cr_2Se_4$		
		$Cu_{1/2}Ga_{1/2}Rh_2S_4$		
		$Li_{1/2}Fe_{1/2}Cr_2O_4$		

† Normal structure. ‡ Inverted structure.

structural distortions take place as the temperature is lowered. The principle changes are small displacements of barium and titanium atoms relative to the oxygen framework, which result in a permanent electric dipole moment characteristic of a ferroelectric phase.

Another structure frequently found amongst mixed oxides is that of spinel. The general formula of spinels corresponds to $A^{2+}OB^{3+}{}_2O_3$ or $A^{2+}B^{3+}{}_2O_4$. Fe, Mg, Zn or Mn are frequently found in the divalent state and Al, Fe, Mn or Cr as the trivalent ions. The unit cell is cubic and **a** varies over the range 8–10·5 Å depending on the chemical constituents; it contains eight formula units AB_2O_4. The oxygen atoms are approximately cubic close-packed, the unit cell containing eight simple c.c.p. cells of oxygen. The cations A are tetrahedrally co-ordinated by oxygen and the cations B are octahedrally surrounded by six oxygen atoms. The structure is most easily visualised as a series of layers as shown in Fig. 7.7.

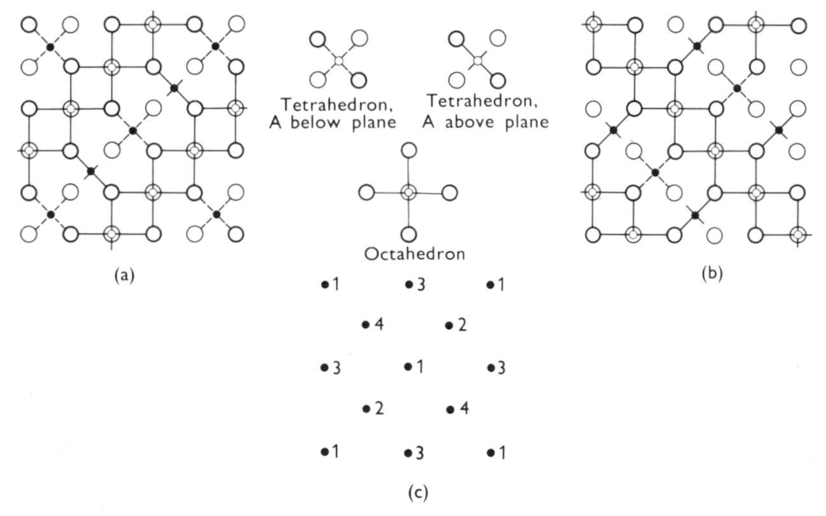

Figure 7.7 (a), (b), and (c), planes of atoms in the spinel structure. For explanation see paragraph 7.6.

The lower layer, (a), has diagonal chains of B octahedra which are linked laterally by A tetrahedra lying alternately above and below the heavily outlined oxygen atoms. In the next layer, (b), the chains form on the other diagonal of the cell. Four such layers make up the complete unit cell and the layer numbers on which the A ions occur are shown in (c): it can be seen that they have the same arrangement as the atoms in diamond (cf. §6.13).

Many examples of spinels occur where the majority cation is not confined entirely to the octahedrally co-ordinated sites as is the case in the *normal spinel* structure. For example, $MgGa_2O_4$ has gallium atoms in the A sites (C.N. = 4) and half the B sites, the remaining B sites being occupied by magnesium. This arrangement is termed the *inverse spinel* structure and the particular example might be more correctly written $Ga(Mg, Ga)O_4$.

Mixed oxides of rare earth elements and the trivalent cations of lighter elements frequently form with the garnet structure. The general formula is $M_5R_3O_{12}$, where R is a rare earth and M may be Fe, Ga, Al etc. Further details of this structure will be deferred to the following section since the prototype, garnet, is a silicate containing isolated SiO_4 tetrahedra.

7.7 Silicates

In the previous section we have noticed the tendency of oxygen atoms to form co-ordination polyhedra around a small cation. Study of compounds exhibiting polymorphism suggests that these co-ordination polyhedra form structural building bricks. This tendency is even more pronounced in the case of the SiO_4 tetrahedron, which forms the basic structural unit of the most important group of rock-forming minerals, comprising some 95% of the Earth's crust. It is convenient to describe this group of structures in terms of the ways in which the SiO_4 tetrahedra are linked within the structure.

Isolated tetrahedra

The structure of olivine $(Mg, Fe)_2SiO_4$ is illustrated in Fig. 7.8 and the isolated SiO_4 tetrahedra can easily be identified. The iron and magnesium cations are octahedrally co-ordinated by the oxygen atoms of two or more SiO_4 groups. In such groups four of the eight negative charges of the oxygen anions are neutralised by the silicon ion, so that each oxygen atom has an unbalanced negative charge of one.

Another frequently occurring structure which contains isolated SiO_4 tetrahedra is that of the garnets. Their general formula is $R_3^{2+}R_2^{3+}(SiO_4)_3$, where R^{2+} is Ca, Mg, Fe or Mn and R^{3+} is Al, Fe or Cr. The unit cell is cubic with a cell edge of some 12 Å and the complete structure is too complicated to illustrate simply. The R^{2+} cations have a C.N. value of 8 and the R^{3+} cations a C.N. value of 6. Each oxygen atom is linked to one R^{3+} ion, one Si atom and two R^{2+} ions. The structure is also exhibited by some mixed oxides in which silicon is replaced by a trivalent rare earth element such as Lu, Yb, Gd etc. and all the remaining cations are also trivalent (see also §7.6).

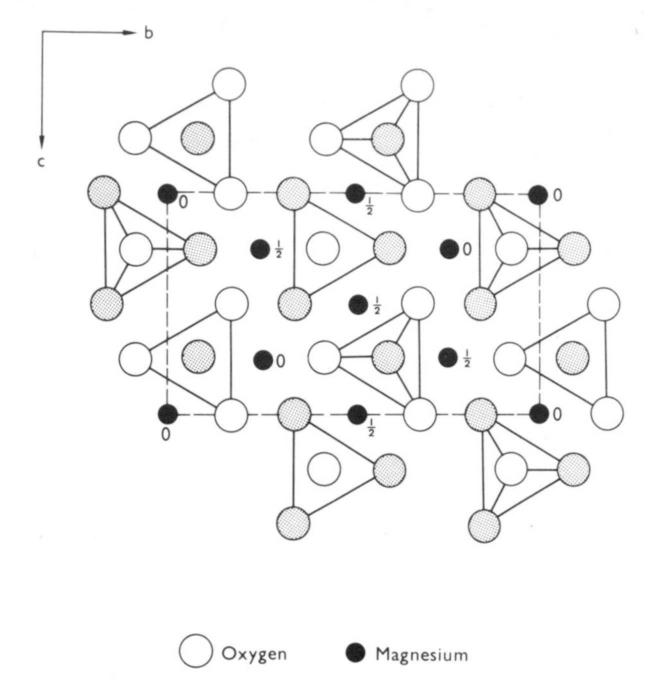

Figure 7.8 The structure of olivine, a simple silicate with isolated SiO_4 tetrahedra. The illustration shows the orthorhombic unit cell. (a = 4·76 Å, b = 10·21 Å, c = 5·99 Å) projected on to (100). The oxygen atoms lie at heights $\frac{1}{4}$ (shaded) and $\frac{3}{4}$ (open). The silicon atoms at the centre of the oxygen tetrahedra are omitted for clarity. The magnesium atoms are in octahedral coordination.

Chained tetrahedra

SiO_4 tetrahedra may be joined together by sharing one or more corners. The simplest case consists of two tetrahedra sharing a single corner to form the Si_2O_7 group found in the structure of the mineral thortveitite, $Sc_2Si_2O_7$. Slightly more complex examples are provided by the structures of benetoite, $BaTiSi_3O_9$, and beryl, $Be_3Al_2Si_6O_{18}$, which contain rings of three and six linked SiO_4 tetrahedra respectively. The Si : O ratio remains constant at 1:3 for any unit in which each tetrahedron shares two corners with other tetrahedra.

The same 1:3 ratio of Si : O occurs in the pyroxene group of minerals in which endless chains of linked SiO_4 tetrahedra are found. A schematic diagram of the structure of one such mineral, diopside, $CaMg(SiO_3)_2$, is given in Fig. 7.9a. Double chains, the basic unit of which is shown in Fig. 7.9b are the basis of the amphiboles having a Si : O ratio of 1:2·75.

Sheets

The double chain illustrated in Fig. 7.9b may obviously be extended to form a two-dimensional sheet by linking to further parallel chains. Such sheets are found in the micas and have a clear connection with the pronounced flaky cleavage of these materials. Mica structures are composed of sandwiches of two sheets arranged with the unshared oxygen atoms pointing inwards to form the co-ordinating atoms of a variety of cations such as Mg and Al. Anions such as OH^- are often found between the sheets as additional co-ordination for the cations. The sandwiches are stacked to form a three-dimensional layered structure and large cations such as potassium may be found between the layers.

Three-dimensional frameworks

If the tetrahedra share all four of their corners to form a framework the Si : O ratio is reduced to 1:2, and consequently such a framework is electrically neutral, without the addition of further cations. Silica, SiO_2, crystallises in four forms, high- and low-temperature quartz, tridymite and cristobalite. The framework of tetrahedral SiO_4 groups is evident in all the structures. The simplest to illustrate is that of cristobalite (Fig. 7.3a) in which the silicon atoms occur in the same positions within the unit cell as in the element (§6.13) and the oxygen atoms occur in positions midway between each pair of silicons. The remaining structures are less symmetrical arrangements of tetrahedra sharing all four corners, and in both modifications of quartz the Si—O—Si bond is bent at the oxygen atom.

7.8 The role of aluminium in the silicates

The radii of Al^{3+} and Si^{4+} for four-fold co-ordination are similar, so that it is possible for aluminium to replace silicon in a tetrahedron of oxygen atoms. The most important consequence of such a substitution is that an average of 1·25 negative charges per oxygen atom now remain unbalanced.

Partial substitution of aluminium for silicon in one of the structures described in the previous section would require an increase in the extra-tetrahedral cation charge, which may be achieved by the substitution of higher valency cations or by the addition of further cations. An example of the latter effect is provided by the micas (§7.7). The number of large

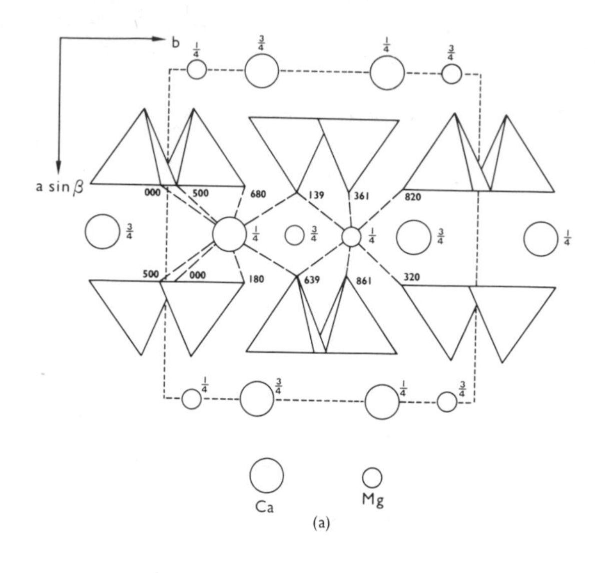

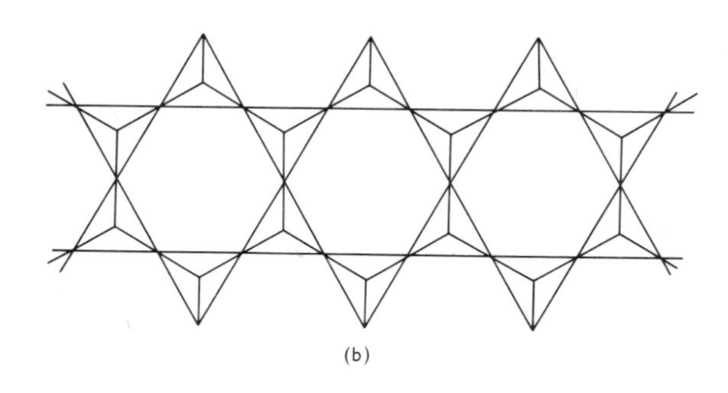

Figure 7.9 (a) A simplified plan of the structure of diopside, $CaMg(SiO_3)_2$, projected down [001]. The monoclinic cell has dimensions a = 9·71 Å, b = 8·89 Å, c = 5·24 Å, β = 74°10′. The chains of linked SiO_4 tetrahedra are seen end on in this projection and the co-ordination of a calcium and a magnesium atom are indicated. (b) Schematic plan view of SiO_4 tetrahedra sharing two and three oxygen atoms to form the double chain found in the amphibole silicates.

cations like potassium which occur between the layers increases as aluminium is substituted for silicon.

The substitution of some aluminium for silicon allows the formation of a number of framework structures which include other cations. The feldspars, e.g. $Na(AlSi_3O_8)$, and the zeolites, e.g. $Na(AlSi_2O_6) \cdot H_2O$, are abundant minerals in which this occurs.

In all formulae of silicates, a distinction must be made between the aluminium which enters into the tetrahedra and the aluminium which is in six co-ordination and plays the same role as the cations magnesium or iron.

7.9 Pauling's principles

Pauling (1929) has discussed the conditions which determine the relative stabilities of ionic structures. Two important new principles emerge. Firstly, we can define an electrostatic bond strength to be associated with any two ions in contact. If a cation A of charge m is co-ordinated by n anions B, then the A–B bond strength is m/n. Pauling's rule states that the sum of the strengths of all bonds to a given anion is equal to its valency. In this way, local electrical neutrality is ensured.

Secondly, Pauling points out that it is energetically favourable to keep similarly charged ions well separated so that the stability of structures decreases as the anion polyhedra share first corners, then edges and finally faces. The lowering of stability is more pronounced if the cation charge is high.

An example of the use of Pauling's rules is afforded by the structure of the silicate benetoite, $BaTiSi_3O_9$(§ 7.7). Each silicon is tetrahedrally co-ordinated by oxygen and each of the titanium and barium atoms are co-ordinated by six oxygens. The bond strengths are therefore:

$$Si^{4+}–O = 1$$

$$Ti^{4+}–O = \tfrac{2}{3}$$

$$Ba^{2+}–O = \tfrac{1}{3}$$

Pauling's rule for local electrical neutrality is satisfied if each oxygen atom is co-ordinated by one silicon, one titanium and one barium atom, which is indeed the case.

7.10 Complex anions

A further deduction from Pauling's rules is that a co-ordination polyhedron in which the electrostatic bond strength is greater than half the charge of each anion cannot link to a similar polyhedron. Such polyhedra will therefore always be identifiable as isolated units in the structure and will form compounds with cations possessing, on average, an equal and opposite charge. In these circumstances it is easy to regard the structures as those of salts derived from an acid $H_xA^{m+}B_n^{((x+m)/n)-}$. No such simple deduction is possible when the polyhedra share members as in the silicates. Of course, any isolated co-ordination polyhedron could be thought of as derived from an acid, though such an acid might be chemically unstable. This concept leads to the names given to some of the mixed oxides referred to in Section 7.6, for example vanadates, molybdates and tungstates.

Table 7.5 Some common complex anions

CO_3, NO_3 planar	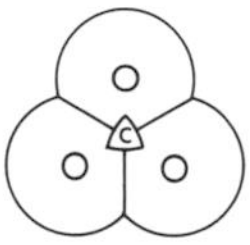 O C O O
SO_4, PO_4, ClO_4 MnO_4, CrO_4 tetrahedral	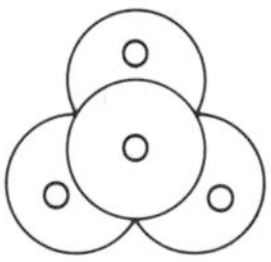 O O O O
CNO, N_3 linear	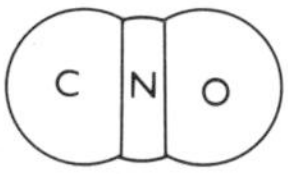 C N O
SO_3, ClO_3, BrO_3 IO_3 pyramidal	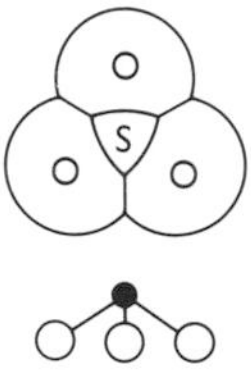 O S O O

The isolated co-ordination group is often more usefully considered as a complex anion. We have seen in Section 7.3 that a high degree of polarisation is associated with a small, highly-charged cation co-ordinated by a larger anion. In these cases the bonding must be thought to be at least partly covalent, and in many cases the complex anions may be understood satisfactorily using a molecular orbital description. Table 7.5 lists some of the more commonly occurring complex anions and illustrates their atomic arrangements. It may be noted that all contain oxygen and hence their electrostatic bond strengths are greater than one.

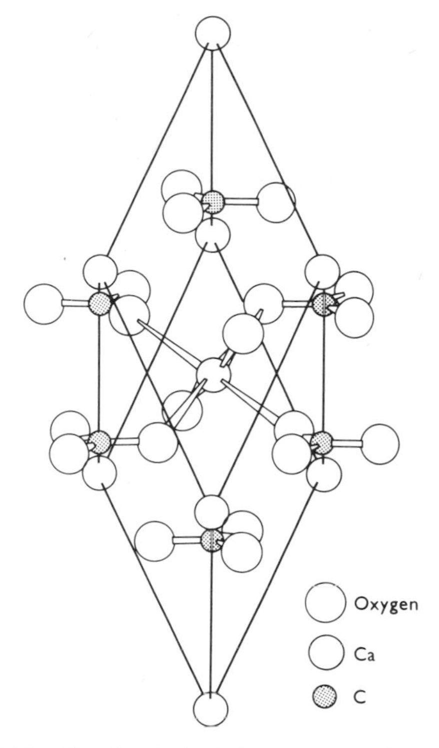

Figure 7.10 The rhombohedral unit cell of calcite, $CaCO_3$.

Many carbonates and nitrates form with the calcite structure, the rhombohedral cell of which is illustrated in Fig. 7.10. The triangular carbonate group lies on the trigonal axis and each cation is octahedrally co-ordinated by oxygen atoms from six different carbonate groups.

The structure of a Group VIII sulphate is shown in Fig. 7.11. Again, the complex ions are easily identified and the cation co-ordination is octahedral.

7.11 Polar effects due to hydrogen

Most of the interesting structural effects introduced by hydrogen have their origin in the small size and consequent high polarising power of the H^+ ion. H^+ is too small to be stably co-ordinated by more than two anions. It does, however, form a complex *cation* in conjunction with nitrogen. The NH_4^+ group differs from the complex anions in as much as the small polarising atoms now surround the anion and cannot be thought of as

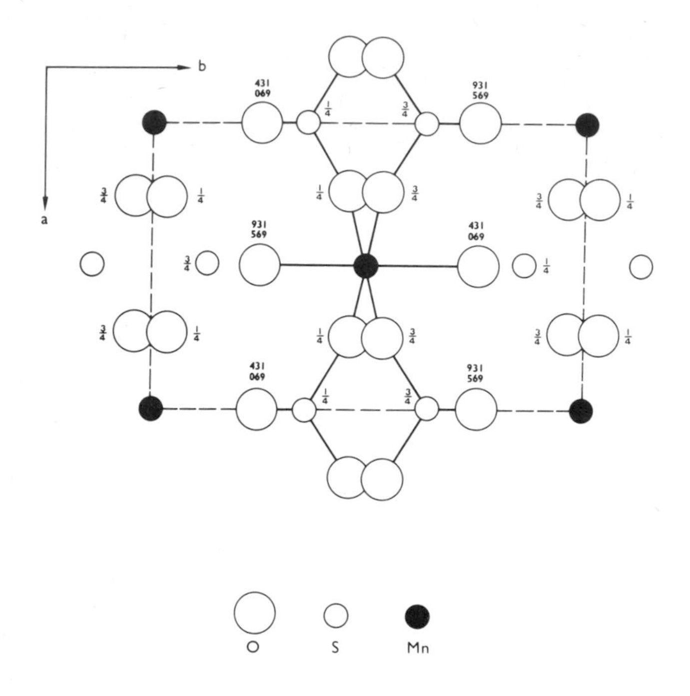

Figure 7.11 The structure of the iron group sulphates. These particular coordinates refer to the structure of manganese sulphate, but $FeSO_4$, $CrVO_4$, $NiSO_4$ and α-$CoSO_4$ are isostructural. The unit cell of $MnSO_4$ is orthorhombic (a = 5·26 Å, b = 8·04 Å, c = 6·85 Å) and a [001] projection is illustrated. The tetrahedral sulphate group can clearly be seen; the transition metal ion is octahedrally coordinated by oxygen atoms belonging to different SO_4^{2-} groups.

co-ordinating it, since they are not in contact with each other. The NH_4^+ group is nearly spherical and occurs in structures in much the same way as the monovalent alkali metals. Many compounds containing ammonium groups undergo structural changes on cooling, and these are associated with this group adopting particular orientations with respect to its environment. To minimise the electrostatic energy resulting from unscreened positive charge the hydrogen atoms are directed towards the co-ordinating anions. The unscreened charge is small, so the associated electrostatic energy is usually comparable with thermal energies allowing libration and free rotation as the temperature is increased.

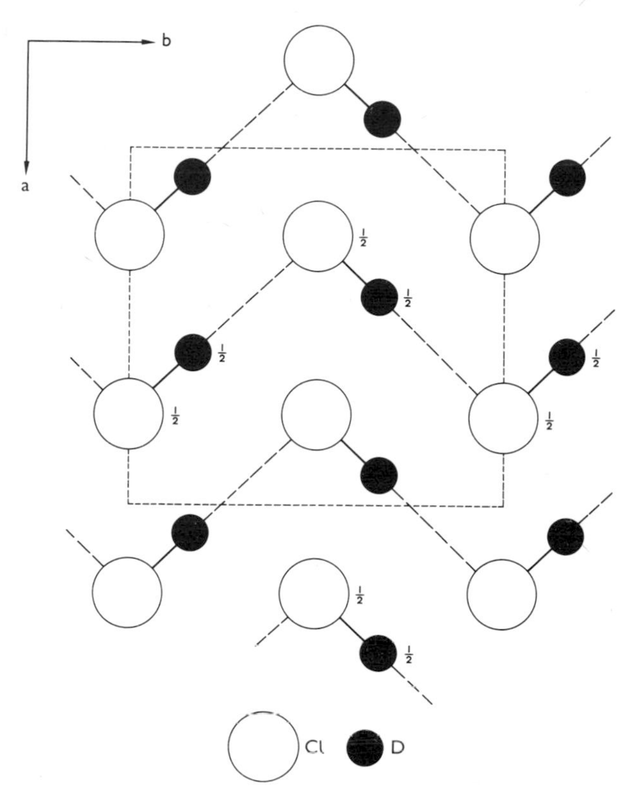

Figure 7.12 The structure of solid DCl at 77·4 K. The orthorhombic phase (a = 5·05 Å, b = 5·37 Å, c = 5·83 Å) undergoes a first order phase change to a cubic structure above 105 K. The asymmetric position of the deuterium atoms between the chlorine atoms is clearly seen; the 'hydrogen' bonds are indicated by broken lines in this projection of the structure onto (001).

The hydroxyl group OH^- is dipolar in character (its dipole moment is 1·66 D, i.e., $1{\cdot}66 \times 3{\cdot}3 \times 10^{-30}$ C m). The group is oriented so as to maximise the distance between its positive (H) end and the co-ordinating cations. The positive end may also contribute to the cohesion of the structure by electrostatic interaction with neighbouring ions (the *hydrogen bond*). A striking example of hydrogen bonding is provided by the structure of HCl as revealed by a neutron-diffraction study of the deuterated compound. The structure is shown in Fig. 7.12 where the asymmetry of the hydrogen position is clearly demonstrated. It can be seen that the structure should possess a permanent dipole moment and hence exhibit ferroelectricity. The asymmetry of hydrogen bonding is also responsible for the ferroelectricity of KH_2PO_4, potassium dihydrogen phosphate (KDP) (see also §7.6).

7.12 Water

Water also illustrates the polarising capability of hydrogen. The structure of ice is illustrated in Fig. 7.13 from which it can be seen that the two hydrogen atoms are not symmetrically placed with respect to the oxygen atom. The charge distribution on the molecule is approximately tetrahedral with two positive and two negative areas.

Many ionic compounds exist in hydrated forms and neutron-diffraction experiments provide direct evidence that the water molecules are oriented so as to be bound into the structure by electrostatic interaction.

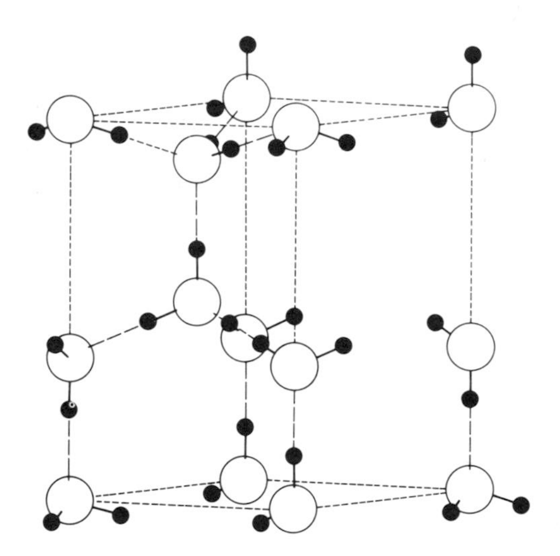

Figure 7.13 The hexagonal structure of ice. The unit cell has dimensions (−50°C) of a = 4·51 Å, c = 7·36 Å and is indicated by dotted lines. Typical hydrogen positions are shown and the hydrogen bonding which results is indicated by broken lines.

7.13 Inert gas compounds

Our final example of polar structure is provided by the inert gases. That these gases can indeed take part in chemical reactions was only fairly recently recognised and the first compounds were prepared in 1962. We

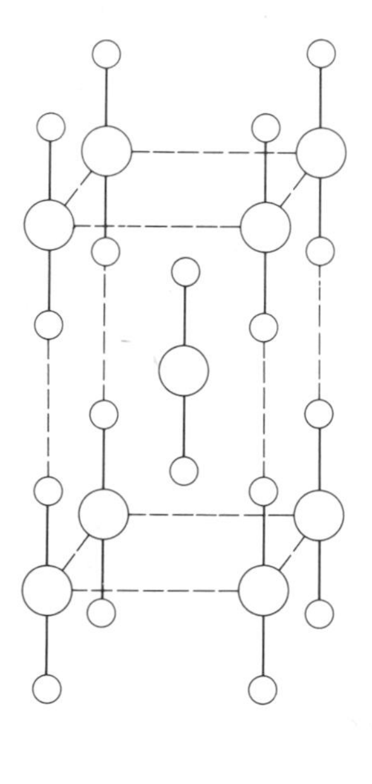

Figure 7.14 The structure of XeF_2 (tetragonal a = 4·32 Å, c = 6·99 Å). The fluorine atoms are 2 Å from the xenon atom with which they are associated. All other interatomic distances are in excess of 3 Å.

have already noted (§6.7) that atoms of the inert gases can polarise each other to some small extent and thereby produce solid phases. The polarisability increases with increasing atomic number, as is shown by the change in melting points of these elements.

Polarisation in ionic compounds may be considered as a partial reversal of the charge separation which changes the covalent to the ionic bond (see §7.3). Similarly, an interaction between a polarisable inert gas atom, such as xenon, and a polarising electron acceptor, such as fluorine, may be expected to introduce a form of covalent bond. However, in this case some charge separation would be involved, i.e. some charge would be transferred from the inert gas to the fluorine.

XeF_2 is a typical inert gas compound, the structure of which is shown in Fig. 7.14; XeF_2 molecules can clearly be seen. A simple molecular orbital model which accounts for this molecule is one in which the valence orbitals are filled except for the $2p_z$ of each fluorine atom and the $5p_z$ of xenon. Four electrons (one from each F and two from Xe) are available to form molecular orbitals. The stability of the molecule is due to the formation of a bonding σ orbital involving p_z functions on all three atoms. The remaining two electrons fill a non-bonding orbital localised mainly on the fluorine atoms.

8

Structures of Binary Alloys

8.1 Introduction

The previous chapter has been concerned with compounds which form between elements with markedly dissimilar electron configurations. Apart from those few light elements which form simple molecules, combinations of elements with similar configurations may be classed as alloys. The study of alloys is simplified by the relatively small size difference between the elements and by the generally non-directional character of the interaction between like atoms. In this situation there is a tendency to form certain simple structures, common to a large number of systems, whose occurrence is determined primarily by size considerations. For example, the extremely simple caesium chloride structure (§7.2) exists in at least 150 binary alloy systems.

Within the framework imposed by the atomic sizes, the structures formed are influenced by the nature of other electronic interactions between the atoms. When the valencies of the two constituents are the same these interactions are essentially similar to those between the pure elements (Chapter 6). When the valencies differ there is always some tendency to charge separation which leads ultimately to the formation of ionic compounds (Chapter 7). In alloy systems the difference in valency is restricted and the polar contribution to cohesion is small.

Our discussion of alloys will be restricted to binary systems since they are sufficient to illustrate the more important principles involved and contain most of the frequently occurring alloy structure types.

8.2 Solid solutions and super-lattices

The simplest alloy systems are those between two metals with the same valency and similar sizes. For example, Au–Ag, Ti–Zr and K–Rb form continuous solid solutions in which the atoms of one element randomly replace those of the other within the framework of a common crystal structure.

As the size difference increases, or if the structures of the constituent elements are different, complete solid solution is not possible. A small difference in atomic size is found to lead to an ordered distribution of one sort of atom in the structure of the other, giving rise to the well-known 'super-lattice' structures. The super-lattices formed in copper gold alloys around the compositions CuAu and Cu_3Au are illustrated in Fig. 8.1a and b.

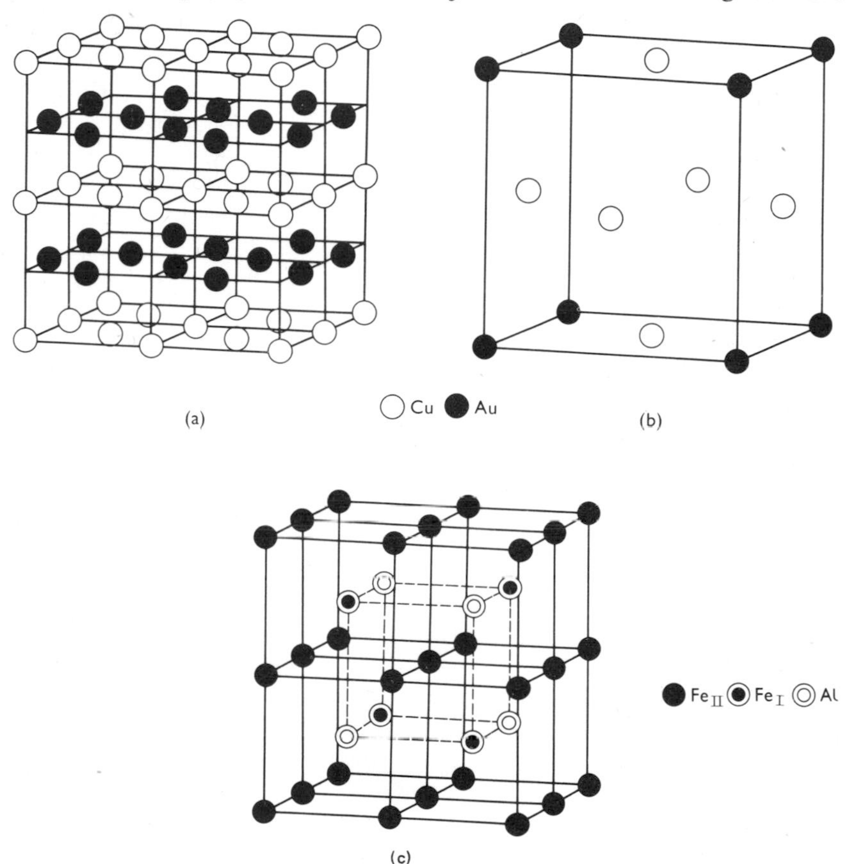

Figure 8.1 The super-lattice structures of (a) CuAu tetragonal a = 3·97 Å, c = 3·67 Å, (b) Cu_3Au cubic a = 3·74 Å, (c) Fe_3Al cubic a = 5·79 Å.

The name 'super-lattice' arose from the study of the X-ray diffraction patterns of such materials in which atomic order changes the crystal lattice and gives rise to extra reflections. In the case of Cu_3Au the lattice changes from F to P, and reflections with mixed odd and even indices appear (cf. § 3.7). An example of a 'super-lattice' which can occur in a body-centred cubic parent structure is that formed by iron–aluminium alloys at the composition Fe_3Al. The atomic arrangement is illustrated in Fig. 8.1c.

Solid solutions are prolific amongst alloys because of the existence of the transition series which provide many pairs of elements of similar size and outer electron configuration. Among the other elements only those in the same group and in adjacent periods satisfy these conditions and can be expected to form extensive solid solutions. Solid solutions may form even between non-metallic elements, for example Si–Ge and Se–Te.

8.3 Alloy packing structures

When the size difference between the constituent atoms of an alloy is more than some 15% the formation of super-lattices becomes energetically unfavourable, and a number of special structures characterised by particular size ratios are formed. Important amongst these are the group of Laves phase structures, of composition AB_2, the basic layer of which is illustrated in Fig. 8.2. The layers may be superposed in several ways to form cubic ($MgCu_2$-type) or hexagonal ($MgZn_2$ and $MgNi_2$) structures, each having the same first neighbour co-ordinations and ideally the same density. This superposition of identical layers to form different structures is analogous to the stacking of close-packed layers to form cubic and hexagonal close-packed structures (§ 6.6).

In the cubic Laves phase the atoms occupy positions fixed by the crystal symmetry and define an ideal size ratio between the constituent atoms A and B of 1·225 : 1. There are over 200 known examples of Laves phases of which about 150 are cubic and most of the remainder are of the $MgZn_2$ type. A salient feature of the Laves phases is the occurrence of a 16-fold (C.N. = 16) polyhedron at the centre of which is an A atom (the larger of the two components). The smaller B atoms are in 12-fold icosahedral co-ordination. The two co-ordination polyhedra are illustrated in Fig. 8.3c and f. The ratio of the atomic radii of the components, r_A/r_B referred to 12-fold co-ordination, is found to vary between 1·05 and 1·68 for known Laves phases. $MgZn_2$-type phases exist over the whole range whereas the cubic phase has not been found at ratios in excess of 1·40. It is clear therefore that size factors alone do not determine the formation of the Laves phases.

Two structures, typified by UNi_5 and $CaCu_5$, occurring at AB_5 stoichiometry are structural variants of the Laves phases. The cubic UNi_5

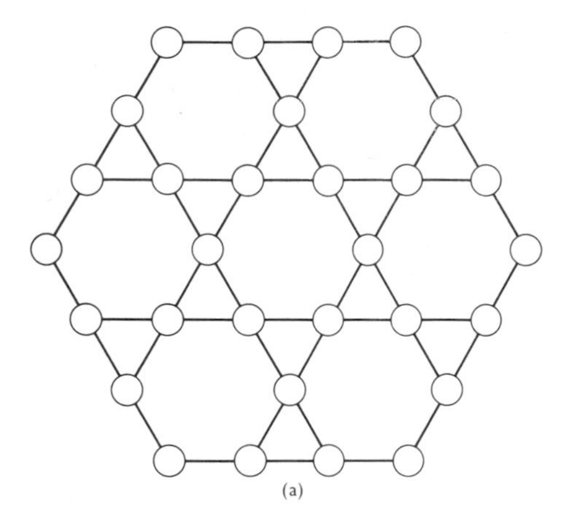

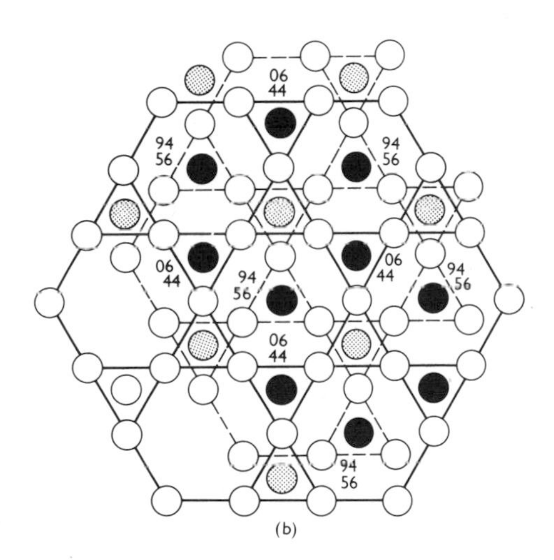

Figure 8.2 (a) The basic layer of smaller (B) atoms found in the Laves phase structures. In (b) the structure of $MgZn_2$ is illustrated, the hexagonal cell (a = 5·15 Å, c = 8·48 Å) being projected on to (001). Two of the layers illustrated in (a) lie at heights $\frac{1}{4}$ (broken lines) and $\frac{3}{4}$ (full lines). Extra B atoms (shaded) occur at height zero in positions with icosahedral coordinates. The A atoms are shown solid and their heights are indicated in hundredths of the c axis. These are in 16-fold co-ordination.

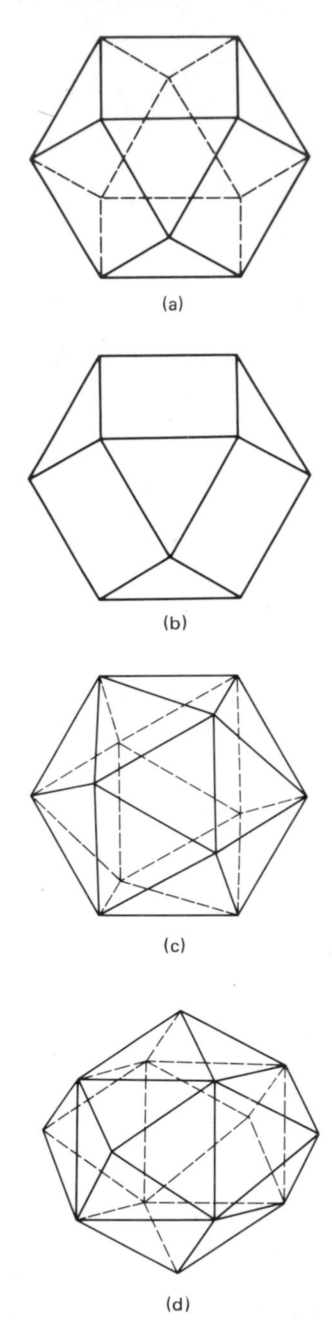

(a)

(b)

(c)

(d)

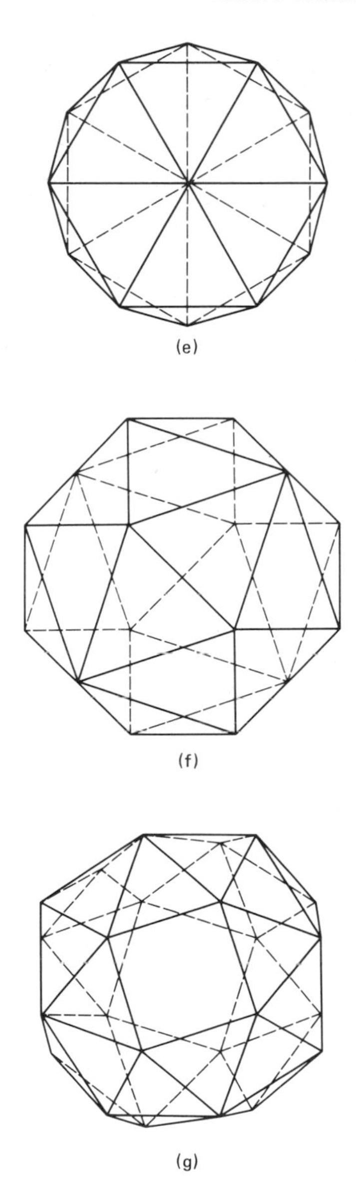

Figure 8.3 Some coordination polyhedra found in intermetallic compounds and pure metals. (a) 12-fold, cubic close packing. (b) 12-fold hexagonal close packing. (c) Icosahedral, Laves phases. (d) 14-fold, body centred cubic. (e) 14-fold, sigma phase. (f) 16-fold, Laves phases. (g) 24-fold, $NaZn_{13}$.

structure is derived from $MgCu_2$ by replacing A atoms by B atoms on four of the atomic sites in the latter structure. The $CaCu_5$ structure is related to that of $MgZn_2$ by a similar replacement combined with some atomic displacements. The ideal radius ratio r_A/r_B for A–A and B–B contact is 1·45 for the UNi_5 structure but all known phases with this structure have $r_A/r_B < 1.3$. The $CaCu_5$-type phases all have $r_A/r_B > 1{\cdot}3$ so that the radius ratio, although not obviously related to the ideal one, seems to be the determining factor in the choice between these two structures.

A structure which forms at the composition AB_2 when A is smaller than B is that typified by $CuAl_2$; the [001] projection of the tetragonal cell is illustrated in Fig. 8.4. The structure is found within the radius ratios r_A/r_B 0·9–0·85; the co-ordination of A is tenfold and that of B is eight.

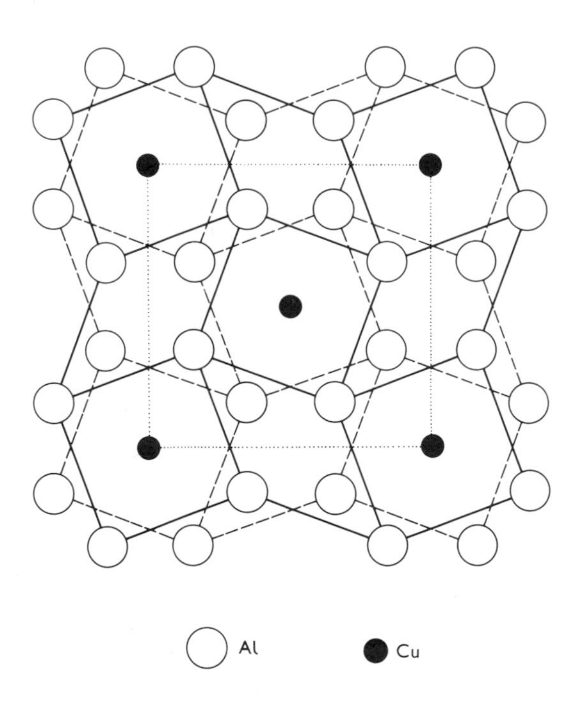

Figure 8.4 The structure of $CuAl_2$ projected down the c axis of the tetragonal unit cell. The copper atoms are at heights $\frac{1}{4}$ and $\frac{3}{4}$ and are superposed in the projection. The bold lines connect aluminium atoms at height $\frac{1}{2}$ and the broken lines aluminium atoms at height zero. The dotted lines indicate the unit cell. (a = 6·05 Å, c = 4·88 Å).

There are six structures with the composition AB_3 which exhibit various stacking schemes of ordered close-packed layers. They are typified by phases $AuCu_3$, $TiNi_3$, $MgCd_3$, $PuAl_3$, $TiCu_3$ and $TiAl_3$. Fig. 8.5a shows the arrangement of atoms in the ordered layers found in the first four structures and 8.5b the layers occuring in $TiCu_3$ and $TiAl_3$. All the stacking variants produce a polyhedron of B atoms about each A atom having a C.N. value of 12.

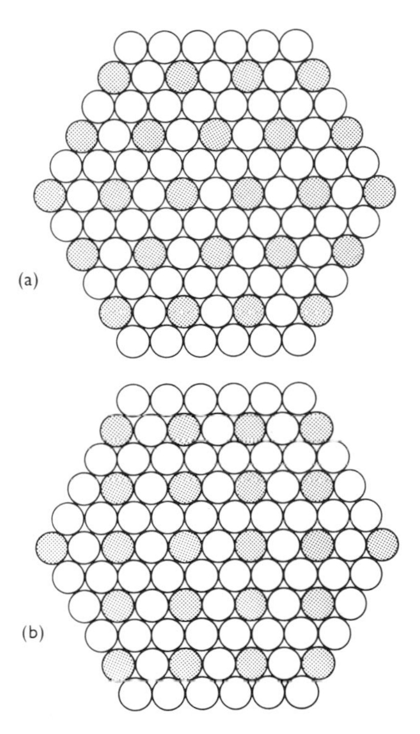

Figure 8.5 Close-packed ordered layers in AB_3 phases. Layer (a) occurs in phases of the $AuCu_3$-type, $TiNi_3$-type, $MgCd_3$-type and $PuAl_3$-type. Layer (b) occurs in $TiCu_3$-type and $TiAl_3$-type phases.

A discussion of alloy packing structures would not be complete without mention of the CsCl structure (§7.2). Clearly this structure is closely related to the b.c.c. structure formed by some true metals (cf. §6.8). It may exist over a wide composition range by super-lattice formation when the size difference is small, or as an approximately stoichiometric compound. The radius ratio can increase up to 1·37 : 1 which corresponds to simultaneous A–A and A–B contacts. The environment of both types of atoms is such that the eight shortest interatomic distances are between unlike atoms. It should therefore not be assumed that the occurrence of this structure necessarily indicates a size-factor alloy, since any affinity between unlike atoms will tend to stabilise the structure.

Size ratios in excesss of 1:4 lead to the formation of structures with large ratios of B to A atoms such as $NaZn_{13}$ and $BaHg_{11}$. They are characterised by high co-ordination of A by B atoms (24 and 22 respectively); the 24-fold polyhedron is illustrated in Fig. 8.3g.

8.4 Binary compounds between elements belonging to neighbouring groups

It is possible to make a rather loose division of the elements into three classes—metals, transition metals and non-metals. Within each of these classes there exist numbers of pairs of closely similar elements whose alloys have been described in the previous two sections. Further types of alloy structures are formed by pairings of less similar atoms within the first two classes.

An important series of compounds exists within the metal systems with one element of noble metal and the other a B sub-group metal. The existence of a systematic tendency to compound formation was pointed out by Hume-Rothery after whom the phases are named. A typical system is provided by Cu–Zn in which four phases occur: c.c.p. solid solution based on copper, CsCl super-lattice, γ-brass structure and h.c.p. solid solution based on zinc. The same four phases occur in other systems of this type but their compositions vary widely between systems. Hume-Rothery pointed out however that the composition at which any particular structure formed corresponded to the same ratio of valence electrons to atoms (e/a ratio). The e/a ratios which characterise the structure types are shown in Table 8.1. Hume-Rothery phases are also found in many systems between transition metals and B sub-group metals, but the e/a ratios are not those expected if

Table 8.1 Electron to atom ratios, e/a, for the Hume-Rothery phases.

Structure	e/a	Examples
b.c.c.	$\frac{3}{2}(\frac{21}{14})$	AuZn, Au_5Sn_4, Cu_3Ga
γ-brass	$\frac{21}{13}$	Ag_5Zn_8, $Cu_{31}Sn_8$, Cu_9Al_4
h.c.p.	$\frac{7}{4}(\frac{21}{12})$	$AgCd_3$, Cu_3Ge, Ag_5Al_2

the normal transition metal valencies are used. Closer agreement can be obtained by assuming that the transition metal contributes no electrons.

The correlation between e/a ratio and structural stability arises from energy associated with filling electronic states within the Brillouin zone (§6.6). Jones (1937) suggests that the transition between the f.c.c. and b.c.c. structures at e/a ~ 1·5 occurs because at this density electrons can be accommodated in the Brillouin zone of the latter structure with a lower energy than in the former. It must be emphasised that the energy of electrons in the Brillouin zone is only one of a number of contributions to the total energy and these other contributions may also be structure-dependent. It is therefore not unusual to find Hume-Rothery phases with non-typical e/a ratios in which the electron/atom ratio is a minor, rather than the dominant, stabilising factor.

8.5 Binary phases of variable composition between two transition metals

We have already described (§8.3) a series of binary alloy packing structures which exist at simple and more or less invariant stoichiometric ratios between the two constituent atoms. In Section 8.4 we have seen that a structure like γ-brass can be stable at widely different compositions in different alloy systems. Further examples of the occurrence of structures capable of existing over wide variations in composition occur in alloys between two transition metals. The three most frequently occurring are the σ-, μ- and χ-phases: Fig. 8.6 illustrates their occurrence (and that of some of the fixed composition phases) in alloy systems of two transition metals. Most of the examples occur between elements one of which, the A type, is a member of the Sc, Ti, V or Cr groups and the other, the B type, is an element of the Mn, Fe, Co, Ni or Cu groups.

All three phases are characterised by high co-ordination and a number of different atomic environments exist in each structure; Table 8.2 summarises this information. The occurrence, compositional limits and atomic ordering in these phases have all been examined in attempts to decide on the relative importance of atomic size and electronic factors in determining the stability of the phases. No single feature is common to all examples and one must suppose that the competing influences decide the final outcome. Several general features are, however, discernable. The σ-phase always has one A-type component (elements in groups to the left of manganese in the Periodic Table) and is found over a fairly small range of radius ratio. X-ray and neutron diffraction studies of the ordering in the σ-phases show that the generally larger A-type atoms occupy the 15-fold sites and the other atoms the 12 and 10 sites. The 14-fold sites may be occupied by either component, except in the rare instances of one component being a non-transition metal, when they are only occupied by transition metal atoms.

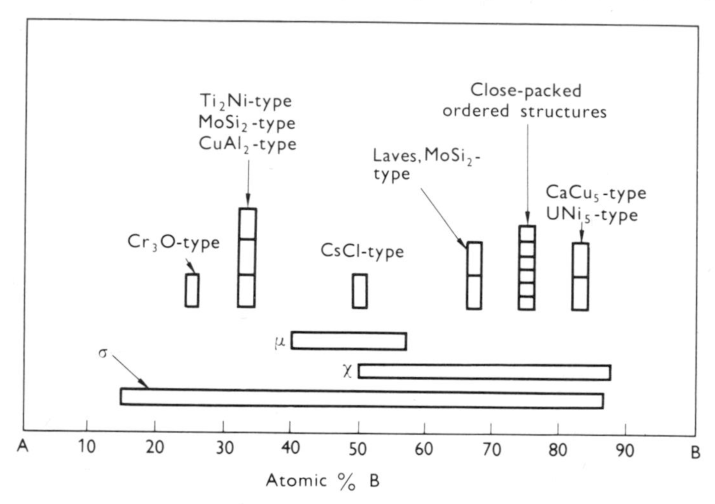

Figure 8.6 The occurrence of some important binary intermediate phases. In transition metal compounds the A component is an element of the Sc, Ti, V or Cr group and the B component is any element of the Mn, Fe, Co, Ni or Cu group.

The composition range for stability of phases with the same A-type atom shifts to higher A concentration as the group number of the A-type atom rises; there is therefore a tendency to retain the same number of d electrons.

The χ-phases exhibit the same tendency. The larger atoms are always of the A type and occupy the 16-fold site; the B type atoms have C.N. = 12 and C.N. = 13. A-type atoms again fill the more highly co-ordinated sites in the μ-phase.

8.6 The Cr_3O structure

A commonly occurring structure at A_3B stoichiometry is that of Cr_3O. This structure was originally ascribed to an allotrope of tungsten which was later shown to be an oxide; for this reason it is sometimes referred to as the β-tungsten structure (*Structurberischt*, type A15).

Table 8.2 The σ-, χ- and μ-phases

Phase	Range of r_A/r_B	No. of atoms per unit cell	No. of d electrons	Atomic site		
				Name	Number	C.N.
σ	0·93–1·15	30	3·6–5·6	A, a	2	12
				B, g	4	15
				C, i	8	14
				D, i	8	12
				E, j	8	14
χ	1·02–1·21	58	4·3–5·0	a	2	16
				c	8	13
				g	24	12
				g	24	12
μ	1·10–1·18	13	5·1–5·6	a	1	12
				h	6	12
				c	2	15
				c	2	16
				c	2	14

The unit cell is cubic and contains two formula units. Each A atom has a C.N. value of 14 and each B atom a value of 12 in a distorted icosahedral arrangement. The A component of the phase is always a transition element but the B component may be another transition element or a B sub-group metal.

The A and C site co-ordination polyhedra for the σ-phase are illustrated in Fig. 8.3c and e, respectively.

In every example the size ratio of the two constituents is within 15% of unity. The structure contains two co-ordinations found in the χ-, μ- and σ-phases discussed in Section 8.5. It does not, however, exhibit the wide variations in composition associated with the χ- and σ-phases, which leads us to suppose that these particular 14- and 12-fold co-ordinations are acceptable only to A and B transition metal elements respectively, irrespective of their relative sizes. Non-transition metal elements do not occupy the highly co-ordinated 14-fold site. Similarly, in the σ-phases which contain Al and Si these elements are again confined to the C.N. = 12 sites.

8.7 Semi-metallic alloys: interstitial compounds

An important class of alloy structures is formed by the transition elements in combination with B sub-group elements. In all of these materials some degree of polar interaction contributes to the stability of the structures. It is convenient to divide this class of materials into two groups depending on the radius ratio of the constituent atoms.

a) Radius ratio in excess of 1·7—interstitial compounds.

These may be thought of in terms of the fitting of the smaller (B) atoms into interstices in the metallic (A) structure. In the c.c.p. structure two types of hole can be distinguished, the eight at positions such as $(\frac{1}{4}\frac{1}{4}\frac{1}{4})$ which are tetrahedrally co-ordinated and the four at positions like $(00\frac{1}{2})$ which are in octahedral co-ordination. The equivalent positions in h.c.p. are four with C.N. values of 4 at positions $(00 \pm \frac{3}{8})(\frac{2}{3}\frac{1}{3} \pm \frac{1}{8})$ and two with C.N. values of 6 at $(\frac{1}{3}\frac{2}{3} \pm \frac{1}{4})$. In both structures the corresponding radius ratios are 4·45 and 2·41 respectively. The b.c.c. structure has only tetrahedral interstices which are rather larger than those in the c.c.p. structure, the critical radius ratio being 3·44.

The most commonly occurring interstitial elements are hydrogen, boron, carbon and nitrogen of which only hydrogen is found in the tetrahedral holes in the c.c.p. and h.c.p. structures (ZrH, TiH and PdH). Carbides and nitrides based on the filling of octahedral holes in the close-packed structure form when r_A/r_B lies in the range 1·69–2·56. If r_A/r_B is less than 1·69, more complicated structures form, e.g. cementite, Fe_3C, with $r_A/r_B = 1{\cdot}64$.

The most common formulae for the interstitial compounds are M_4X, M_2X, MX and MX_2 where M is the transition metal element. The type M_2X is particularly liable to form wide ranges of solid solution corresponding to the filling up of the interstices. Interstitial compounds usually retain much of the electrical conductivity of the parent metal, but the extent of the interaction between the interstitial and the host atoms may be judged from the increased melting points of such compounds as TiC (3150 °C), ZrC (3530 °C) and TaC (3900 °C) compared to that of the transition metal itself.

b) Radius ratio less than 1·7.

A great variety of structure types exist for materials in which the radius ratio is smaller than 1·7. Some of the most common types are shown by Fe_3P, Fe_3C and Fe_2P. These structures are usually described in terms of co-ordination polyhedra around transition metal atoms, the most common configuration being a triangular prism. These structures are characterised by a narrow composition range, but otherwise have properties more nearly those of the metal than the metalloid constituent.

8.8 Nickel arsenide and related structures

The structure of nickel arsenide is illustrated in Fig. 8.7. It can be seen that the metalloid atoms form a close-packed hexagonal arrangement and that the metal sites are octahedrally co-ordinated by metalloid atoms.

Phases possessing this structure frequently exhibit a variable composition Me_xB, with x from 0·8 to 2·0. For $x > 1$ a second set of metal sites at $\pm(\frac{1}{3}\frac{2}{3}\frac{3}{4})$ are occupied. When $x = 0{\cdot}5$ only half the original octahedral sites are filled and the structure is then identical to the CdI_2 structure (§7.4). $CoTe_2$ and $NiTe_2$ are stable at this composition.

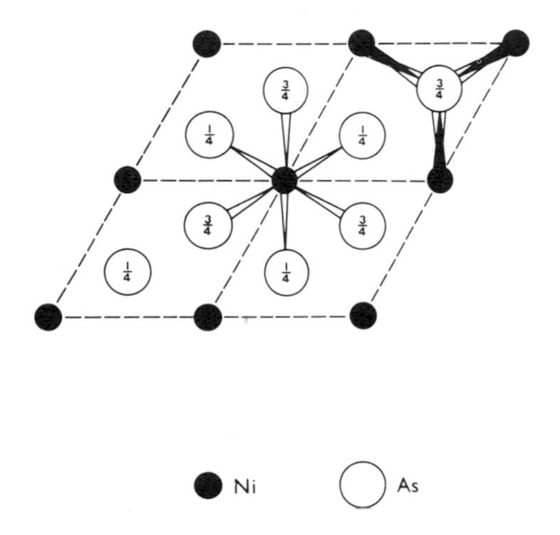

Figure 8.7 Four hexagonal unit cells (a = 3·62 Å, c = 5·03 Å) of the NiAs structure projected on to (001). The nickel atoms are superimposed at heights 0 and $\frac{1}{2}$. The nickel atoms are in octahedral co-ordination whereas the six nickel neighbours of each atom form the corners of a trigonal prism.

A number of more complicated structures exist which are closely related to NiAs, for example Mn_5Si_3 and MnP. These types of structure have limited ranges of composition and often form when a phase possessing the nickel arsenide structure at high temperatures is annealed. An example is provided by CrS_{1+x} which decomposes to Cr_7S_8, Cr_5S_6 and Cr_3S_4, all three phases having different structures related to that of nickel arsenide.

8.9 Alloys exhibiting typically ionic structures

We have noted in the previous section that using half the octahedral holes in the NiAs structure produces the typically ionic structure of CdI_2. Similarly, filling all the octahedral interstitial holes in the c.c.p. structure results in the familiar NaCl structure (§7.2). Other examples of intermetallic compounds possessing ionic structures are given in Table 8.3.

Table 8.3 Alloys exhibiting typically ionic structures.

Phase	Ionic structure	
NiS_2, CoS_2	fluorite, CaF_2	(§7.4)
RuSi	CsCl	(§7.2)
MnSe	wurtzite	(§7.2)
CrH	zinc blende	(§7.2)
ZrS_2, $NiSe_2$	CdI_2	(§7.4)

The structural feature common to binary ionic compounds is that any atom is co-ordinated exclusively by atoms of the other type and this arises from the strong interaction between the unlike atoms. Such an interaction is clearly responsible for stabilising such structures amongst alloys. The degree to which charge separation actually occurs varies widely and may be inferred from properties like electrical conductivity and colour. Clearly these materials form a bridge between the variable composition alloys considered earlier in this chapter and the strictly stoichiometric ionic compounds of Chapter 7.

9

Conclusion

It has not been our intention in this short book to provide a comprehensive catalogue of structures, but to illustrate the principles underlying structural stability by considering a number of commonly occurring or important structures. We have not mentioned many molecular structures since the configurations of their molecules are dictated by chemical principles and the arrangement of molecules within the crystalline material has less significance. The structural principles of organic materials are described in many textbooks and an introduction to the subject may be found in *Organic Chemical Crystallography* by A. I. Kitaigorodskii (Consultants Bureau, New York, 1957) and *Diffraction of X-rays by Proteins, Nucleic Acids and Viruses* by H. R. Wilson (Arnold, London, 1966). However, because of the lack of common structure types amongst these materials it is unlikely that the structure of a particular new compound can be found in these books.

A complete index of organic compounds whose structures are known is kept by the Crystallographic Data Centre, University Chemical Laboratory, Cambridge. Tabulated data on crystal structures can also be found in a growing number of books and these data are increasingly being transferred to computer-based data banks. In addition, a number of international journals specialise in reporting new structure determinations. Some useful books and journals are listed in the first part of the Bibliography.

The crystal structure of a material is an important factor governing its physical properties, including those dependent on defects in the structure, and its role is emphasised in some of the other books in this series. The solid-state physicist may therefore need to find out whether the structure of a particular material is known by using the references referred to above. It may therefore be useful to indicate how this can be done for particular cases by giving two examples.

9.1 A search for the structure of an alloy material

The phase diagram of the Co–In system has recently been reported (Schöbel and Stadelmaier, 1970). Two intermetallic phases are formed of which one, $CoIn_3$, has a small tetragonal cell with dimensions $a = 6{\cdot}82$ Å and $c = 3{\cdot}53$ Å. No further structural details are given.

In the case of an unknown alloy structure it is always worth making a careful search through the known structure types, since the likely structure may be deduced in this way. The most easily used reference book for alloy structures is *A Handbook of Lattice Spacings and Structures of Metals and Alloys*, Volume 2, by W. B. Pearson. Table 6 of this book gives the critical data for alloy materials published up to 1965. It contains no reference to $CoIn_3$. In Schubert's book on binary alloys, classification is done by reference to the Group numbers of the constituent elements. $CoIn_3$ is therefore a T_9B_3 alloy (T = transition metal, B = B sub-group metal). The appropriate table however contains no reference to $CoIn_3$.

It may be possible to infer the likely structure from a knowledge of the cell constants and the known composition. The first step is to find the number of formula units in the unit cell. Since in this case the density is not given, this cannot be done directly. However, it is possible to estimate the atomic volume per formula unit by reference to Fig. 6.2, which gives the atomic radii. The values found are $r_{In} = 1{\cdot}62$ Å and $r_{Co} = 1{\cdot}25$ Å and these lead to a volume of $61{\cdot}5$ Å^3/$CoIn_3$. Intermetallic compounds are generally well packed, so we may assume that the packing fraction will be near to that for the close-packed metals, viz. 0·740 (§6.7). The estimate for the volume occupied by each formula unit is therefore $83{\cdot}1$ Å^3, whereas the cell dimensions give a unit-cell volume close to twice this figure, $164{\cdot}2$ Å^3. We may therefore conclude that there are two formula units per cell.

Table 4 of Pearson's book gives a classification of the structures of metals and alloys according to Bravais lattice and the number of atoms in the crystallographic unit cell. In the case of $CoIn_3$ the Bravais lattice is tetragonal and must be either *P* or *I* (§1.4); the number of atoms in the cell is 8. There is no entry under t*P*8 and a single prototype, $TiAl_3$, at the correct composition against t*I*8. Reference to Table 6 of the book gives the cell dimensions of $TiAl_3$ as $a = 3{\cdot}84$ Å and $c = 8{\cdot}58$ Å, which are very different from those of $CoIn_3$.

Another way of attacking the problem is to look for references to structures of alloys between chemically similar elements at the composition 3 : 1. Schubert's book contains a table giving the structure types which occur at particular compositions. Reference to the table at 25% reveals a chemically related phase $CoGa_3$, which is tetragonal with cell dimensions $a = 6{\cdot}25$ Å and $c/a = 1{\cdot}03_5$ and 16 atoms per unit cell. The a dimension agrees well with the $CoIn_3$ phase, allowing for the larger size of indium. However,

$CoIn_3$ has a c/a ratio of 0·51₉, half that of $CoGa_3$. If the two structures are related, we expect to find that the upper and lower halves of the larger cell are very similar.

Figure 9.1 is a plane of the structure of $CoGa_3$ projected on (001). The unit cell can indeed be divided into two halves, which are identical with respect to the gallium atoms. They differ in that the cobalt atoms in one

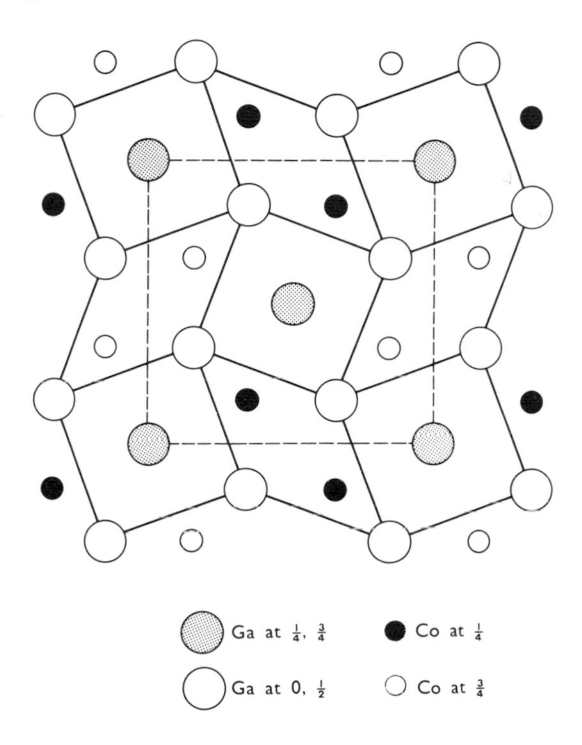

Figure 9.1 The structure of $CoGa_3$ projected down the c axis of the tetragonal unit cell. The bold lines connect gallium atoms in layers at heights 0 and $\frac{1}{2}$. The structure is closely related to that of $CuAl_2$ (Figure 8.4). The broken lines indicate the unit cell (a = 6·26 Å, c = 6·48 Å).

are rotated 90° with respect to those in the other. A unit cell with c one-half that in $CoGa_3$ would result if the cobalt atoms were randomly distributed over the eight positions. Such a model might well serve as a trial structure from which a complete structure refinement could be made.

9.2 The use of structural information to interpret physical properties

Molybdenum disulphide is used in greases to improve their lubricating properties. We are familiar with the structural basis for a similar property in graphite and we may therefore expect to be able to explain this quality in MoS_2 by reference to its structure.

Structure Reports issues cumulative indices, 1940–1950 and 1950–1960. The earlier volume tells us that work on MoS_2 is reported in Volume 8, p. 137 and Volume 11, p. 41. The first of these reports refers us to *Structurberieht*, Volume 1, p. 164 for the crystal structure. This source gives us the following information:

MoS_2 *C*7 type. Hexagonal unit cell

$a = 3{\cdot}15$ Å $\quad c = 12{\cdot}3_0$ Å

Mo at $(\frac{1}{3}\frac{2}{3}\frac{1}{4})$
S at $(\frac{1}{3}\frac{2}{3}\frac{1}{4} + u)$ with $u = 0{\cdot}379$

Two molecules per unit cell, space group D^4_{6h}, $P6_3/mmc$.

International Tables for Crystallography, Volume 1, give these special positions as

(Mo) 2 c $\pm(\frac{1}{3}\frac{2}{3}\frac{1}{4})$
(S) 4 f $\pm(\frac{1}{3}\frac{2}{3}z)$, $\pm(\frac{2}{3}\frac{1}{3}\frac{1}{2} + z)$
i.e. $z = u + \frac{1}{4}$

Figure 9.2a shows the [0001] projection of the unit cell constructed from these data. The interatomic distances are as given in Table 9.1.

Evaluating the Mo–S distance gives a result of 2·35 Å. Figure 9.2b shows that the structure is indeed layered and that the shortest distance between atoms in different layers is 3·65 Å (S–S), which is considerably longer than twice the atomic radius for sulphur found from Table 6.2 (1·05 Å). We may therefore conclude that the lubricating properties of MoS_2 arise from the pronounced (0001) cleavage which results from the weak inter-layer binding.

9.3 The unknown structure

In the examples given in the preceding two sections structural information has been available in the larger compilations. These works are always, by their nature, several years out of date so that failure to find the required information does not necessarily mean that it is not available. In such instances, a search through recent volumes of *Chemical Abstracts*

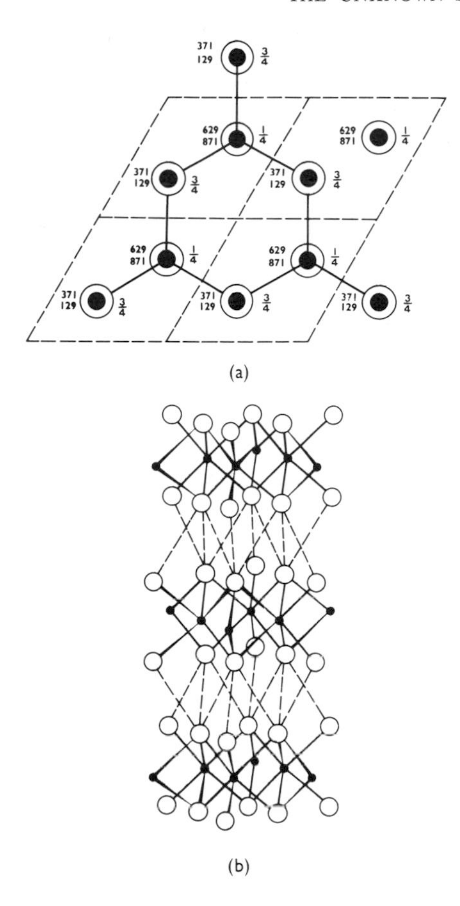

Figure 9.2 The structure of MoS_2. (a) Four hexagonal unit cells (a = 3·15 Å, c = 12·30 Å) projected on to (001). The full circles represent Mo atoms at heights $\frac{1}{4}$ and $\frac{3}{4}$. The open circles represent sulphur atoms at the fractional heights 0·129, 0·371, 0·629 and 0·871. The full lines show the co-ordination of molybdenum atoms at height $\frac{1}{4}$ by six sulphur atoms at the corners of a trigonal prism. (b) Successive layers of MoS_2 are weakly bonded together as indicated by the broken lines.

Table 9.1 Interatomic distances in molybdenum disulphide

Atom	Neighbour	C.N.	Distance
Mo	S	6	$[a\frac{2}{3} + c^2(\frac{3}{4} - z)^2]^{1/2}$
S	Mo	3	$[a\frac{2}{3} + c^2(\frac{3}{4} - z)^2]^{1/2}$

may give references to relevant papers. In addition, the indices of recent volumes of *Acta Crystallographica* and *Zeitschrift für Kristallografie* should be searched. Should information not be found in any of these places, it is probable that the crystal structure is unknown.

The first stage in the examination of an unknown structure should be a determination of the Bravais lattice and cell constants and it may well be (see §9.1) that these may then show that the structure is related to some known type. If this is not the case, the determination of the crystal structure may be a complex task, the complete description of which is beyond the scope of this book. For the interested reader, there are several specialised books on structure determination listed in the second part of the Bibliography.

Appendix

The Fourier Transform

The function $F(u)$ defined by the integral equation

$$f(x) = \int_{-\infty}^{\infty} F(u) \exp(-2\pi iux)\, du$$

is called the *Fourier Transform* of $f(x)$ and is given by

$$F(u) = \int_{-\infty}^{\infty} f(x) \exp(2\pi iux)\, dx$$

These relationships may be written as

$$F(u) = T[f(x)] \text{ and } f(x) = T^{-1}[F(u)]$$

By the operation T we mean multiplication by $\exp(-2\pi iux)$ followed by integration; by T^{-1} we mean multiplication by $\exp(2\pi iux)$ prior to integration.

We note that $T^2[f(x)] = f(-x) = T[f(u)]$ etc. For even functions $f(x)$

$$F(u) = 2\int_{0}^{\infty} f(x)(\cos 2\pi ux)\, dx$$

and for odd functions $f(x)$

$$F(u) = 2\int_{0}^{\infty} f(x) \sin(2\pi ux)\, dx$$

Let us now derive the Fourier transforms for three simple cases:

(i) a delta function. $\delta(r)$ satisfies $\delta(r) = 0$ except at $r = 0$ and $\int \delta(r)\, dv = 1$

$$F(u) = \int \delta(r)\, e^0\, dv = \int \delta(r)\, dv = 1.$$

We see that the Fourier transform of a δ function is a constant distribution.

(ii) a rectangular distribution in one dimension:

$$f(x) = a \text{ for } |x| < b$$
$$= 0 \text{ for } |x| > b$$

$$F(u) = \int_{-b}^{+b} a \exp(2\pi iux)\, dx$$

$$= \frac{a}{2\pi iu} \left[\exp(2\pi iux)\right]_{-b}^{+b}$$

$$= 2a \frac{\sin 2\pi bu}{2\pi bu}$$

The form of $F(u)$ is illustrated in Fig. 3.1*a*.

(iii) a Gaussian in three dimensions

$$\rho(r) = \exp(-pr^2)$$

$$F(u) = \left(\frac{\pi}{p}\right)^{3/2} \exp(-\pi u^2/p^2)$$

that is, the Fourier transform is another Gaussian function.

Tables 2.5.3.A, B, C and D in Volume 2 of the *International Tables for X-ray Crystallography* list many of the mathematical properties of Fourier transforms and the Fourier transforms of commonly occurring functions. In particular, we note that the Fourier transform of the sum of two functions is the sum of the Fourier transforms of the functions. Another important relationship concerns the Fourier transform of the product of two functions, the *convolution* formulae.

$$f(x)g(x) = T^{-1}\left[\int_{-\infty}^{\infty} F(u-\eta)\, G(\eta)\, d\eta\right]$$

$$\text{or } T^{-1}\left[\int_{-\infty}^{\infty} F(\eta)\, G(u-\eta)\, d\eta\right]$$

and

$$F(u)G(u) = T\left[\int_{-\infty}^{\infty} f(\eta)\, g(x-\eta)\, d\eta\right]$$

$$\text{or } T\left[\int_{-\infty}^{\infty} f(x-\eta)\, g(\eta)\, d\eta\right]$$

The simplest example is the convolution of a function with a δ function. Since the Fourier transform of the delta function is a uniform distribution, the result is simply the Fourier transform of the function.

References

Barett, C. S. 1956. *Acta Cryst.,* **9**, 671.
Bauer, E. 1969. *Techniques of Metals Research* Vol II pt 2, Ch 16, Wiley, New York.
Bloch, F. 1928. *Z. Phys.,* **52**, 555.
Brown, P. J. 1957. *Acta Cryst.,* **10**, 133.
Buerger, M. J. 1960. *Crystal Structure Analysis,* John Wiley and Sons Inc, New York and London.
Caspar, D. L. D. 1956. *Nature,* **177**, 475.
Cohen, M. U. 1935. *Rev. scient. Instrum.,* **6**, 68.
Debye, P. 1920. *Z. Phys.,* **21**, 178.
Von Halben, H. and Prieswork, P. 1936. *C.r. hebd. Séanc. Acad. Sci.,* Paris, **263**, 73.
Henry, N. F. M., Lipson, H. and Wooster, W. A. 1951. *The Interpretation of X-ray Diffraction Photographs,* Macmillan, London.
Hirsch, P. B., Howie, A., Nicholson, R. B., Pashley, D. W. and Whelan, M. J. 1965. *Electron Microscopy of Thin Crystals,* Butterworth's, London.
Hume-Rothery, W. and Raynor, G. V. 1938. *Phil. Mag.,* **26**, 129.
International tables for X-ray crystallography Vol I, 1952, Vol II, 1959, Vol III, 1962. Kynoch Press, Birmingham.
Ito, T. 1950. *X-ray Studies of Polymorphism,* Maruzen Co. Ltd, Tokyo.
James, R. W. 1948. *The Optical Properties of the Diffraction of X-rays,* Bell, London.
Jones, H. 1937. *Proc. phys. Soc.,* **49**, 250.
Klug, A. Finch, J. T. and Franklin, R. E. 1957. *Biochim. biophys. Acta,* **25**, 242.
Koster, G. F. 1955. *Phys. Rev.,* **98**, 901.
London, F. 1930. *Z. Phys.,* **63**, 245.
Mott, N. F. and Massey, H. W. 1949. *The Theory of Atomic Collisions.* Oxford University Press, Oxford.
Pauling, L. 1929. *J. Am. chem. Soc.,* **51**, 1010.
Tessman, J. R., Khan, A. H. and Shockley, W. 1953. *Phys. Rev.,* **92**, 890.
Schöbel, J-D. and Stadelmaier, H. H. 1970. *Z. Metallk.,* **61**, 342.
Vainstein, B. K. and Pinsker, Z. G. 1958. *Kristallografiya, Soviet Phys. Crystallogr.,* **3**, 358.
Wilson, R. R. and Levinger, T. S. 1964. *A. Rev. nucl. Sci.,* **14**, 135.

Bibliography

Books and Journals containing useful structural data

BOOKS

Crystal Structures. R. W. G. Wykoff, Interscience, New York. Loose leaf format 1948, 1951, 1953, 1957, 1958, 1959, and 1960. Republished in book form by the Interscience Division of John Wiley, New York. Volume I, 1963, Volume II, 1964, Volume III, 1965.

A Handbook of Lattice Spacings of Metals and Alloys. W. B. Pearson, Pergamon Press, London. Volume 1, 1958, Volume 2, 1967.

Kristallstrukturen zweikomponentiger Phasen. K. Shubert, Springer Verlag, Berlin, 1961.

Strukturbericht. Edited by P. P. Ewald et al., Academische Verlagsgesellschaft mbH, Leipzig and photo lithoprint reproduction by Edwards Bros. Inc. Michigan, 1943. Volumes 1-7 covering structural investigations during the years 1913 to 1938.

Structure Reports. Edited by A. J. C. Wilson *et al.*, Oosthoek, Utrecht. Volumes 8 onwards cover structural investigations from 1940. Volume 14 contains a cumulative index 1940-1950 and volume 25 one for 1951-1960.

Tables of Interatomic Distances and Configuration in Molecules and Ions compiled by H. J. M. Bowen *et al.* Special publication No. 11, The Chemical Society, London, 1958. Continued and extended as a series of volumes entitled *Molecular Structures and Dimensions,* Oosthoek, Utrecht. Volume 1 deals with general organic crystal structures and volume 2 with complexes, organo-metals and metalloids. Both volumes cover the years 1935-1969. Further volumes will be published.

JOURNALS

Acta Chemica Scandinavica.
Acta Crystallographica, Section B, Structural Crystallography and Crystal Chemistry.
Journal American Chemical Society.
Kristallografiya SSSR (Translated as *Soviet Physics Crystallography*).
Zeitschrift für Kristallographie.

Some books on the subject of X-ray structure determination

Buerger, M. J. 1959. *Vector Space and its Application in Crystal Structure Investigation,* John Wiley and Sons Inc., New York and London.

Hauptman, H. and Karle, J. 1953. *Solution of the Phase Problem,* American Crystallographic Association ACA Monograph No. 3.

Kitaigorodskii, A. I. 1961. *The Theory of Crystal Structure Analysis* (translated by Harker, D. and Harker, K. from the Russian) Consultants Bureau Inc., New York.

Lipson, H. and Cochran, W. 1966. *The Determination of Crystal Structures* (*The Crystalline State,* volume 3), Bell, London.

Lipson, H. and Taylor, C. A. 1958. *Fourier Transforms and X-ray Diffraction,* Bell, London.

Nyburg, S. C. 1961. *X-ray Analysis of Organic Structures,* Academic Press, London.

Stout, G. H. and Jensen, L. H. 1968. *X-ray Structure Determination,* Macmillan, New York.

Woolfson, M. M. 1961. *Direct Methods in Crystallography,* Oxford University Press, Oxford.

Pepinsky, R. et al. (editors) 1961. *Computing Methods and the Phase Problem in X-ray Crystal Analysis,* Pergamon Press, Oxford.

Index

Chemical compounds are indexed by formulae in which the constituent elements are alphabetically ordered. Where appropriate, they are also indexed in full: e.g. caesium chloride, ClCs.